An Introduction to Contemporary Remote Sensing

About the Author

Dr. Qihao Weng is a professor of geography and the director of the Center for Urban and Environmental Change at Indiana State University.  He is also a guest/adjunct professor at Peking University, Beijing Normal University, Wuhan University, and Fujian Normal University, and a guest research scientist at the Beijing Meteorological Bureau. His research focuses on remote sensing and GIS analysis of urban ecological and environmental systems, land-use and land-cover change, urbanization impacts, and human-environment interactions. Dr. Weng is the author of more than 130 peer-reviewed journal articles and other publications and 4 books. He has received a number of awards, including the Outstanding Contributions Award in Remote Sensing from the American Association of Geographers (AAG) Remote Sensing Specialty Group (2011), the Erdas Award for Best Scientific Paper in Remote Sensing from the American Society for Photogrammetry and Remote Sensing (ASPRS) (2010), a NASA Senior Fellowship (2009), and the Theodore Dreiser Distinguished Research Award from Indiana State University (2006). Dr. Weng has worked extensively with optical and thermal remote sensing data, and his projects have been funded by NSF, NASA, USGS, USAID, NOAA, the National Geographic Society, and the Indiana Department of Natural Resources. At Indiana State, he teaches courses on remote sensing, digital image processing, remote sensing–GIS integration, and GIS and environmental modeling, and has mentored 10 doctoral and 7 masters students. Dr. Weng has served as a national director of ASPRS (2007–2010), chair of the AAG China Geography Specialty Group (2010–2011), and a member of the U.S. Department of Energy's Cool Roofs Roadmap and Strategy Review Panel (2010). Currently, he is an associate editor of the *ISPRS Journal of Photogrammetry and Remote Sensing* and the series editor for the *Taylor & Francis Series in Remote Sensing Applications*. In addition, Dr. Weng is the task leader for the Group on Earth Observations' SB-04, Global Urban Observation and Information (2012–2015).

An Introduction to Contemporary Remote Sensing

Qihao Weng, Ph.D.

New York   Chicago   San Francisco
Lisbon   London   Madrid   Mexico City
Milan   New Delhi   San Juan
Seoul   Singapore   Sydney   Toronto

Copyright © 2012 by The McGraw-Hill Companies, Inc. All rights reserved. Printed in China. Except as permitted under the United States Copyright Act of 1976, no part of this publication may be reproduced or distributed in any form or by any means, or stored in a data base or retrieval system, without the prior written permission of the publisher.

3 4 5 6 7 8 9 0 CTP/CTP 1 7 6 5

ISBN 978-0-07-174011-1
MHID 0-07-174011-2

Sponsoring Editor
Michael Penn

Editing Supervisor
Stephen M. Smith

Production Supervisor
Richard C. Ruzycka

Acquisitions Coordinator
Bridget L. Thoreson

Project Manager
Anupriya Tyagi,
Cenveo Publisher Services

Copy Editor
Susan Fox-Greenberg

Proofreader
Paul Sobel

Indexer
Robert Swanson

Art Director, Cover
Jeff Weeks

Composition
Cenveo Publisher Services

McGraw-Hill books are available at special quantity discounts to use as premiums and sales promotions, or for use in corporate training programs. To contact a representative, please e-mail us at bulksales@mcgraw-hill.com.

This book is printed on acid-free paper.

Contents

Preface

At the turn of the 21st century, we have witnessed great advances in remote sensing and imaging science. Commercial satellites, such as QuickBird and IKONOS, acquire imagery at a spatial resolution previously only possible with aerial platforms, but these satellites have advantages over aerial imageries including their capacity for synoptic coverage, inherently digital format, short revisit time, and the capability to produce stereo image pairs conveniently for high-accuracy 3D mapping thanks to their flexible pointing mechanism. Hyperspectral imaging affords the potential for detailed identification of materials and better estimates of their abundance in the Earth's surface, enabling the use of remote sensing data collection to replace data collection that was formerly limited to laboratory testing or expensive field surveys. Lidar (light detection and ranging) technology can provide high-accuracy height and other geometric information for urban structures and vegetation. In addition, radar technology has been re-inventing itself since the 1990s. This is greatly due to the increase of spaceborne radar programs, such as the launch of European Radar Satellite (ERS-1) in 1991, Japanese Earth Resources satellite (J-ERS) in 1992, ERS-2 and Radarsat in 1995, and the Advanced Land Observation Satellite (ALOS) in 2006. The Shuttle Radar Technology Mission (SRTM) started to gather data for digital elevation in 2000, providing over 80% of global land coverage. In the meantime, we see more diverse applications using the technique of radar interferometry. These technologies are not isolated at all. In fact, their integrated uses with more established aerial photography and multispectral remote sensing techniques have been the mainstream of current remote sensing research and applications.

With the advent of the new sensor technology, the reinvention of "old" technology, and more capable computational techniques, the field of remote sensing and Earth observation is rapidly gaining, or re-gaining, interest in the geospatial technology community, governments, industries, and the general public. Today, remote sensing has become an essential tool for understanding the Earth and managing human–Earth interactions. Global environmental problems have become unprecedentedly important in the 21st century. Thus, there is a rapidly growing need for remote sensing and Earth observation technology that will enable monitoring of the world's natural resources and environments, managing exposure to natural and man-made risks and more frequently occurring disasters, and helping the sustainability and productivity of natural and human ecosystems. Driven by societal needs and technology advances, many municipal government agencies have started to collect remote sensing image data, and jointly used archived GIS data sets for civic and environmental applications. Richardson (2004, *Nature*, 427:6972, Jan. 22) assured that "the U.S. Department of Labor identified geotechnology as one of the three most important emerging and evolving fields, along with nanotechnology and biotechnology." It should be noted that the integration of Internet technology with remote sensing imaging science and GIS have led to the emergence of geo-referenced information over the Web, such as Google Earth and Virtual Globe. These new geo-referenced "worlds," in conjunction with GPS, mobile mapping, and modern telecommunication technologies, have sparked much interest in the public for remote sensing and imaging science, especially among young students.

On college campuses, we see an ever-increasing number of undergraduate students requesting to take remote sensing or related science/technology classes in recent years. To meet the needs of non-major students, many universities have opened general education (service-type) classes in remote sensing and other geospatial technologies. However, instructors have struggled to find a suitable textbook for their "first" remote sensing class because textbooks in the market are written mainly for upper-division undergraduate students and graduate students, who have decided to major in geospatial science or related subjects. The textbooks suitable for the lower-division students are few and most were written before the emergence of contemporary remote sensing and imaging, Internet, and modern telecommunication technologies. This is the main reason I wrote this book.

Given the breadth of contemporary remote sensing technology and methods, it is necessary to provide an outline here for students to navigate the book. In Chap. 1, we define remote sensing, briefly review its history, and illustrate a few examples of its wide applications. Further, we discuss how other modern geospatial technologies—GIS, GPS, mobile mapping, and Google Earth—are all intertwined with remote sensing for good. Chapter 2 introduces the principle of electromagnetic radiation and how it is used in remote sensing. Chapter 3 discusses remote sensing data recording methods, which are closely related to the characteristics of remotely sensed data. Chapters 4 and 5 deal with airborne remote sensing. Chapter 4 focuses on cameras, films, and filters, as well as the methods for aerial photo interpretation, whereas Chap. 5 examines the principles of photogrammetry, with more emphasis on quantitative analysis of aerial photographs, and ends with an introduction to orthophotography. Chapter 6 presents non-photographic sensing systems, including electro-optical, thermal infrared, passive microwave and imaging radar, and Lidar sensors. Chapter 7 reviews remote sensing platforms and orbital characteristics, and introduces major satellite programs for Earth's resources, meteorological, oceanographic, and commercial uses. Digital image processing is a sub-field of remote sensing that has undergone a rapid development recently. Chapter 8 reviews most commonly employed methods in digital image processing, while also extending our discussion into the most up-to-date techniques in hyperspectral remote sensing and object-based image analysis. Chapter 9 describes thermal-infrared remote sensing principles and thermal image analysis and interpretation. The final chapter, Chap. 10, introduces the methods of data collection and analysis of two active sensing systems, radar and Lidar.

I started to teach an introductory remote sensing class (general education type) at Indiana State University 10 years ago. The class started from a very low enrollment course, offered one session per semester, and grew to an exceedingly popular general education class, which often has to be offered in multiple sessions every fall, spring, and summer terms. To meet the needs of additional students, I created a WebCt version of the course in 2002 and transformed into the Blackboard version in 2005. All sessions may reach the full capacity in a short time; so from time to time, students must be placed on a waiting list. My students' concern over a textbook for that class was my initial motivation for writing a text. In this sense, I appreciate all comments and suggestions received from my former students, some of whom I have never seen face-to-face. Those are students in my long-distance education classes, who may reside in Florida, New Mexico, or Hawaii.

I further thank colleagues in the United States, China, and Brazil who expressed a strong interest in such a textbook. I hope that the publication of this book provides a favorable choice for their remote sensing classes. Moreover, I wish to express my sincere appreciation to Taisuke Soda, former acquisitions editor at McGraw-Hill Professional. Mr. Soda convinced me that the international market for such a textbook would be huge because many developing countries, like China, India, and Brazil, have placed a strategic emphasis on remote sensing, Earth observation, and related geospatial technologies. Finally, I am indebted to my family for their enduring love and support. My wife, Jessica, helped me prepare numerous figures, diagrams, charts, and tables in this book, while exercising her artistic talent. My daughter and older son read this preface, and provided good suggestions on English writing. Writing a textbook can be boring at times. I thank my younger son for bringing all sorts of joy over the course of writing, which enabled me to remain spirited.

It is my hope that the publication of this book will provide a good choice for undergraduate students searching for a textbook. This text may also be useful for various remote sensing workshops or used as a reference book for graduate students, professionals, researchers, and alike in academics, government, and industries who wish to get updated information on recent developments in remote sensing. If you have any comments, suggestions, or critiques, please contact me at qhweng@gmail.com.

Qihao Weng, Ph.D.

CHAPTER 1

The Field of Remote Sensing

1.1 Introduction

Remote sensing refers to the activities of recording/observing/ perceiving (sensing) objects or events at far away (remote) places. In a more restricted sense, remote sensing refers to the science and technology of acquiring information about the Earth's surface (land and ocean) and atmosphere using sensors onboard airborne (aircraft, balloons) or spaceborne (satellites, space shuttles) platforms. Depending on the scope, remote sensing may be broken down into:

1. Satellite remote sensing—when satellite platforms are used

2. Photography and photogrammetry—when photographs are used to capture visible light

3. Thermal remote sensing—when the thermal infrared portion of the spectrum is used

4. Radar remote sensing—when microwave wavelengths are used

5. Lidar remote sensing—when laser pulses are transmitted toward the ground and the distance between the sensor and the ground is measured based on the return time of each pulse

Remote sensing has now been integrated with other modern geospatial technologies such as geographic information system, global positioning system, mobile mapping, and Google Earth.

1.2 How Does Remote Sensing Work?

We perceive the surrounding world through our five senses. Some senses (for example, touch and taste) require contact of our sensing organs with the objects. However, much information about our

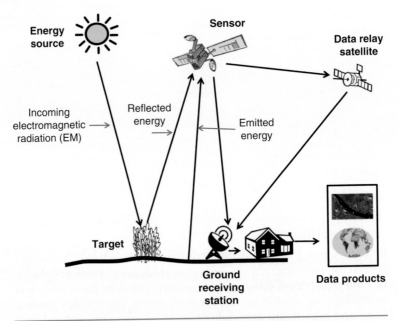

FIGURE 1.1   An illustration of a remote sensing process.

surroundings through the senses of sight and hearing do not require close contact between our organs and the external objects. In this sense, we are performing remote sensing all the time. Generally, remote sensing refers to the activities of recording/observing/ perceiving (sensing) objects or events at far away (remote) places. The sensors (for example, special types of cameras and digital scanners) can be installed in airplanes, satellites, or space shuttles to take pictures of the objects or events in the Earth's surface. Therefore, the sensors are not in direct contact with the objects or events being observed. The information needs a physical carrier to travel from the objects to the sensors through an intervening medium. Electromagnetic radiation (solar radiation) is normally used as an information carrier in remote sensing. The output of a remote sensing system is usually an image (digital picture) representing the objects/events being observed. A further step of image analysis and interpretation is often required in order to extract useful information from the image. Figure 1.1 shows the process of remote sensing.

1.3   Characteristics of Remote Sensing Digital Images

All sensors (sensing systems) detect and record energy "signals" from Earth surface features or from the atmosphere. Familiar examples of remote sensing systems include aerial cameras and video recorders.

Sensing systems that are more complex include electronic scanners, linear/area arrays, laser scanning systems, etc. Data collected by these remote sensing systems can be either in analog format, for example, hardcopy aerial photography or video data, or in digital format, such as a matrix of "brightness values" corresponded to the average radiance (energy) measured within an image pixel. Digital remote sensing images may be directly input into a computer for use, and analog data can be converted into digital format by scanning. Modern computer technology, as well as Internet technology, has transformed remote sensing into a digital information technology.

Although you engage in remote sensing every day with your eyes as the sensor, remote sensors are different from your own "sensor" (eyes). Human eyes can only see visible light, but remote sensors can "see" beyond the visible light and can detect both visible and non-visible reflected and emitted radiation. Some remote sensors can even "see" through clouds and soil surface. Each remote sensor is unique with regard to what portions of the electromagnetic spectrum (known as "bands") it detects, and is often designed to "see" specific features on the Earth's surface. A well-known sensor is MODIS (moderate resolution imaging spectrometer), on both NASA's Terra and Aqua missions and its follow-on. It is engineered to provide comprehensive data about land, ocean, and atmospheric processes simultaneously. MODIS has a 2-day repeat global coverage in 36 spectral bands.

Remote sensors are different from the regular cameras (regardless if they are analog or digital) we use in our daily life. Remote sensors carried by airplanes, satellites, space shuttles, and other platforms provide a unique "bird's-eye" view (view from the top) of the Earth's surface. This synoptic view allows us to examine and interpret objects or events simultaneously on a large area and to determine their spatial relationships. In contrast, regular cameras can only provide a ground-level view of the objects or events (i.e., the "side view"). In fact, most remote sensing satellites are equipped with sensors looking down to the Earth. They are the "eyes in the sky," and provide constant observation of the globe as they go around in predictable orbits. Depending on the types of applications and funds available, different spatial resolutions of remote sensors are employed for different missions. Generally speaking, at the local scale, high spatial-resolution imagery is more effective. At the regional scale, medium spatial resolution data are the most frequently used. At the continental or global scale, coarse spatial resolution data are most suitable.

Many environmental phenomena constantly change over time, such as vegetation, weather, forest fires, volcanoes, and so on. The revisit time is an important consideration in environmental monitoring.

For example, because vegetation grows according to seasonal and annual phenological cycles, it is crucial to obtain anniversary or near-anniversary images to detect changes in vegetation. Many weather sensors have a much shorter re-visit time: Geostationary Operational Environmental Satellite (GOES) is every 30 minutes, NOAA AVHRR (National Oceanic and Atmospheric Administration Advanced Very High Resolution Radiometer) local area coverage is half a day, and Meteosat Second Generation is every 15 minutes.

1.4   A Brief History of Remote Sensing

The technology of remote sensing gradually evolved into a scientific subject after World War II. Its early development was driven mainly by military uses. Later, remotely sensed data became widely applied for civic usages. Melesse, Weng, Thenkabail, and Senay (2007) identified eight distinct eras in remote sensing development. Some eras overlap in time, but are unique in terms of technology, data utilization, and applications in science. A discussion of these periods follows.

Airborne remote sensing era: The airborne remote sensing era evolved during World War I and II (Colwell, 1983; Avery and Berlin, 1992). During that time, remote sensing was used mainly for the purposes of surveying, reconnaissance, mapping, and military surveillance.

Rudimentary satellite remote sensing era: The spaceborne remote sensing era began with the launch of "test-of-concept" satellites such as Sputnik 1 from Russia and Explorer 1 from the United States at the end of the 1950s (Devine, 1993; House, Gruber, Hunt, and Mecherikunnel, 1986). This was soon followed by the first meteorological satellite, called Television and Infrared Observational Satellite-1 (TIROS-1), by the United States in the late 1950s (House et al., 1986).

Spy satellite era: During the peak of the Cold War, spy satellites such as Corona (Dwayne, Logsdon, and Latell, 1988) were widely employed. Data were collected, almost exclusively for military purposes. The data were not digital, but were produced as hard copies. However, the spin-off of the remote sensing developed for military purposes during the previous eras spilled over to mapping applications and slowly into environmental and natural resources applications.

Meteorological satellite era: The early meteorological satellite sensors consisted of geo-synchronous GOES and polar-orbiting NOAA AVHRR (Kramer, 2002). This was an era when data started being available in digital format and were analyzed using exclusive computers. This was also an era when global coverage became realistic and environmental applications practical.

Landsat era: The Landsat era began with the launch of Landsat-1 (then called Earth Resources Technology Satellite) in 1972 carrying a Return Beam Vidicon (RBV) camera and Multi-spectral Scanner (MSS) sensor. This was followed by other path-finding Landsat satellites 2 through 3, and Landsat 4 and 5, which carried Thematic Mapper (TM) in addition to MSS. After the Landsat-6 failed during launch, Landsat 7 improved its sensor technology and carried the Enhanced Thematic Mapper (ETM+) sensor. The Landsat era also has equally good sun-synchronous land satellites such as Système Pour l'Observation de la Terre (SPOT) of France and Indian Remote Sensing Satellite (IRS) of India (Jensen, 2007). These satellites have moderate resolution, nominally 10 to 80 m, and have global coverage potential. At this resolution, only Landsat is currently gathering data with global coverage. This is, by far, the most significant era that kick-started truly environmental applications of remote sensing data locally, regionally, and globally.

Earth observing system era: The Earth observing system (EOS) era (Bailey, Lauer, and Carneggie, 2001; Stoney, 2005) began with the launch of the Terra satellite in 1999 and has brought in global coverage, frequent repeat coverage, high level of processing (e.g., georectified, at-satellite reflectance), and easy and mostly free access to data. The Terra/Aqua satellites carrying sensors such as MODIS have daily re-visit and various processed data. Applications of remote sensing data have become widespread and applications have multiplied. The availability of the processed data in terms of products such as leaf area index (LAI) and land use/land cover (LULC) have become routine. Currently, MODIS provides more than 40 products. In addition, active spaceborne remote sensing sensors using radar technology have become prominent due to the launch of European Radar Satellite (ERS), Japanese Earth Resources satellite (JERS), Radarsat, and Advanced Land Observation Satellite (ALOS). The Shuttle Radar Technology Mission (SRTM) started to gather data for digital elevation in 2000, providing almost 80% of global land coverage.

New millennium era: The new millennium era (Bailey et al., 2001) refers to highly advanced "test-of-concept" satellites sent into orbit around the same time as the EOS era, but the concepts and technologies are different. These are satellites and sensors for the next generation. These include Earth Observing-1, carrying the first spaceborne hyperspectral sensor, Hyperion. The idea of advanced land imager (ALI) as a cheaper, technologically better replacement for Landsat is also very attractive to users.

Commercial satellite era: This era began at the end of the last millennium and the beginning of this millennium (Stoney, 2005). A number of innovations should be noted. First, data are collected with very high resolution, typically less than 5 m. IKONOS and Quickbird satellites exemplify this. Second, a revolutionary means of data collection emerges. This is typified by a Rapideye satellite constellation of five satellites, having almost daily coverage of any spot on Earth at 6.5-m resolution in five spectral bands. Third, microsatellites are introduced. For example, disaster-monitoring constellations are designed and launched by Surrey Satellite Technology Ltd. for Turkey, Nigeria, China, USGS, UK, and others. Finally, the innovation of Google Earth provides rapid data access of very high resolution images for any part of the world through streaming technology, which makes it easy for even a non-specialist to search and use remote sensing data.

1.5   Examples of Remote Sensing Applications

The range of remote sensing applications includes archaeology, agriculture, cartography, civil engineering, meteorology and climatology, coastal studies, emergency response, forestry, geology, geographic information systems, hazards, land use and land cover, natural disasters, oceanography, water resources, military, and so on. Most recently, with the advent of high spatial-resolution imagery and more capable techniques, urban and other civic applications of remote sensing are rapidly gaining interest in the remote sensing community and beyond. Five examples are discussed in the following.

1.5.1   Crop Irrigation in Kansas

Figure 1.2 shows an interesting pattern of crop fields found in Finney County in southwestern Kansas, United States. Resembling a work of modern art, variegated green crop circles cover what was once short-grass prairie in southwestern Kansas. The most common crops in this region are corn, wheat, and sorghum. Each of the crops was at a different point of development when the Advanced Spaceborne Thermal Emission and Reflection Radiometer (ASTER) captured this image on June 24, 2001, displaying various shades of green and yellow. Healthy, growing crops are green. Corn would grow into leafy stalks by late June. Sorghum grew more slowly and was much smaller, and therefore appeared paler in the image. Wheat was a brilliant gold, as harvest would occur in June. Fields of brown were recently harvested and plowed under or lay fallow for the year.

Like crops throughout large sections of the U.S. Midwest, these crops are fed partly by water from the Ogallala Aquifer, a giant layer of underground water. Although the aquifer is a reliable source of water for irrigated cropland, there are some concerns that it could eventually run dry. Most of the water in the aquifer is "fossil water" from the last ice age. One conservation measure is using water more wisely so less is drawn out of the aquifer. Farmers in this region have adopted a quite efficient irrigation method called "central pivot irrigation." Central pivot irrigation draws water out of a single well in

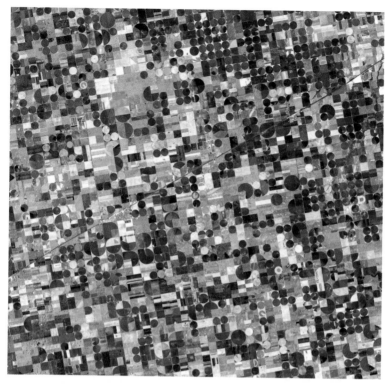

FIGURE 1.2   Crop circles in southwestern Kansas, U.S.A. (*Images courtesy of NASA/GSFC/METI/ERSDAC/JAROS and U.S./Japan ASTER Science Team.*)

the center of the field. Long pipes perched on wheels rotate around the pivot, showering the crops with water. Because the water falls directly on the crops instead of being shot into the air as does with traditional sprinklers, less water is lost to evaporation and more goes to nourishing the growing plants. Central pivot irrigation also creates perfectly circular fields, as seen in this image (Fig. 1.2). The fields are typically 800 and 1600 m (0.5 and 1 mi) in diameter.

1.5.2   Urban Sprawl in Las Vegas, Nevada

Landsat images in 1984 (TM image) and 2002 (Landsat ETM+ image) were selected for a study of urban sprawl in Las Vegas, Nevada, United States (Xian, 2007). Image pixels were classified as "urban," when impervious surface area (ISA, another name for the built-up area) was equal to or greater than 10%, whereas pixels of less than 10% ISA were classified as non-urban. Furthermore, urban development density was defined as 10 to 40% ISA for low-density urban, 41 to 60% ISA for medium-density urban, and above 60% ISA for high-density urban. Figure 1.3 shows the changes in the spatial extent

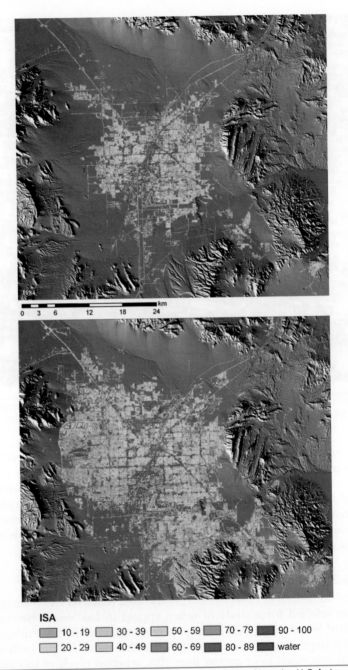

ISA

10 - 19	30 - 39	50 - 59	70 - 79	90 - 100
20 - 29	40 - 49	60 - 69	80 - 89	water

FIGURE 1.3   Urban development density in Las Vegas, Nevada, U.S.A. in 1984 and 2002 (Xian, 2007). (*Used with permission from CRC Press.*)

in urban land use for the Las Vegas Valley between 1984 and 2002. Urban land use expanded in almost all directions in the valley. During the 1980s and early 1990s, most medium- to high-percentage ISA were located in the downtown and Las Vegas strip areas. More recently, high-percentage ISA has expanded to the southeast and northwest portions of Las Vegas. Consistent urban development accomplished with high- and medium-density residential areas, as well as commercial and business centers, enlarged the urban land use boundary far away from the 1984 urban boundary. The areal extent of urban land use was approximately 290 km^2 in 1984 and increased to about 620 km^2 in 2002, representing an increase of 113% during the observed period. Urban land use counted for approximately 10% of total land in the entire mapping area in 1984 and changed to 22% in 2002. To investigate growth rates for different urban development densities, pixels were regrouped into nine categories from 10 to 100% in every interval of 10% imperviousness. It is found that Categories 5–7 (50–70% ISA) contained the largest portion of total ISA in both 1984 and 2002. This indicates that the largest increase of urban land use is in medium- to high-urban development densities that include most single- to two-housing units.

1.5.3   Remote Sensing of Aerosols

This case study was developed by Wong and Nichol (2011). Aerosols are solid or liquid airborne particulates of variable composition, which reside in stratified layers of the atmosphere. Aerosol scattering is more difficult to deal with than Rayleigh scattering because of the great variability in the nature and concentration of aerosol particles in the atmosphere. Generally, aerosols are defined as atmospheric particles with sizes between about 0.1 μm and 10 μm, although the sizes of condensation nuclei are typically about 0.01 μm. Under normal conditions, most of the atmospheric aerosol exists in the troposphere. Natural sources such as dust storms, desert and soil erosion, biogenic emissions, forest and grassland fires, and sea spray account for approximately 90% of aerosols, and the rest result from anthropogenic activity. The background (natural) tropospheric aerosols are temporally and spatially variable. The study of aerosols is important because of their effects on the Earth's radiation budget, climate change, atmospheric conditions, and human health.

Due to inadequacy of fixed air quality stations, there is increasing interest in satellite sensors for synoptic measurement of atmospheric turbidity based on aerosol optical thickness. The task of aerosol monitoring at the local and city scale, over densely urbanized regions, is more challenging because higher spatial resolution is required, and the generally bright surfaces are indistinguishable from the aerosol signal. MODIS is a sensor aboard the Terra and Aqua satellites. With 36 wavebands at 250 m, 500 m, and 1 km resolution, MODIS can be used for atmospheric, oceanic, and land studies at both global and local scales. MODIS also provides specific products such as atmospheric aerosols, ocean color, land cover maps, and fire products. A network of AERONET stations distributed around the world is currently used for calibration of the MODIS satellite observations for aerosols.

Hong Kong, a city with a service-based economy located in southeast China, has suffered serious air pollution over the last decade (Fig. 1.4). During the long winter dry season, air masses are mainly northeasterly bringing continental pollution into the Pearl River Delta region and Hong Kong. The consequent effects on visibility and health due to continuous bad air have appeared gradually since 2000. The Hong Kong Environmental Protection Department reported that an increase of 10 $\mu g/m^3$ in the concentration of NO_x, SO_2, respiratory suspended particulate (RSP), and ozone may cause associated diseases such as respiratory diseases and chronic pulmonary and cardiovascular heart diseases to increase by 0.2% to 3.9%, respectively. The low visibility observed in Hong Kong caused by air pollutants also affects marine and air

(a) A snapshot of polluted Hong Kong

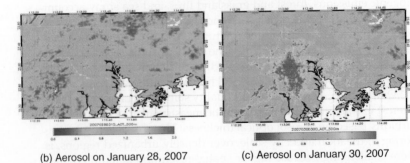

(b) Aerosol on January 28, 2007   (c) Aerosol on January 30, 2007

FIGURE 1.4   Aerosol optical thickness at 550 nm and 500 m resolution over Hong Kong and the Pearl River Delta region of China on Jan 28 and Jan 30, 2007. (*Wong and Nichol, 2011. Used with permission from CRC Press.*)

navigation, and is damaging Hong Kong as a tourist destination. Wong and Nichol (2011) demonstrated the feasibility of using 500-m aerosol optical thickness (AOT) for mapping urban anthropogenic emissions, monitoring changes in regional aerosols, and pinpointing biomass-burning locations. In Fig. 1.4, AOT at 550 nm is mapped with 500-m resolution over Hong Kong and the Pearl River Delta region of China on January 28 and January 30, 2007. The strong contrast in the concentration of aerosols was attributed to the northeasterly wind. The pollutants were progressively accumulated as wind speeds decreased from 4 ms^{-1} on January 28, 2007 to 2 ms^{-1} on January 29, 2007 and 1 ms^{-1} on January 30, 2007.

1.5.4 Monitoring Forest Fire with High Temporal Resolution Satellite Imagery

Rajasekar (2005) conducted research to monitor forest fire using high temporal resolution remote sensing imagery, that is, the Meteosat imagery. As an illustration, a diurnal cycle of fire activity in a selected region of Portugal on July 28, 2004 was mapped and analyzed. The images used were the Meteosat thermal infrared band 4. The temperature in Kelvin for the possible areas of fire was found to be relatively high as compared to their surroundings, but the regions under thermal activity were not readily visible. This was due to low spatial resolution of the imagery. A mathematical method (i.e., Kernel convolution) was developed to characterize the hearth of the fire as an object in space. Objects were then extracted and tracked over time automatically (Fig. 1.5). From Fig. 1.5, we can see that fire objects are characterized as peaks with respect to the background. Fire objects

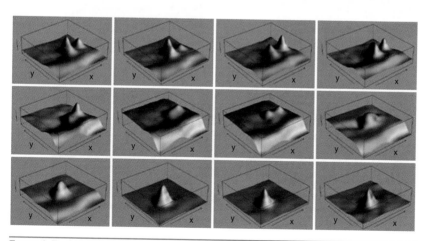

FIGURE 1.5   Modeled fire objects at different times: 00:00, 00:15, 02:30, 05:00, 07:30, 10:00, 12:30, 15:00, 17:30, 20:00, 22:30, and 23:45 h (*Rajasekar, 2005*).

changed in intensity and location with respect to time. Two objects of fire were visible at the beginning of the day (00:00 h). These fires were seen merging later into one huge fire (15:00 h). Background temperature slowly increased, reaching the maximum around noon (10:00–12:30) and then gradually decreased.

1.5.5   Remote Sensing Application in the Recovery at the World Trade Center

On September 11, 2001, terrorists used hijacked planes as missiles to attack the World Trade Center in New York City. Recovery at the site went on for months. On September 14, 2001, the New York State Office for Technology (OFT) asked Earth Data to assist in this recovery at the World Trade Center. The mission, jointly developed by OFT and Earth Data, was to conduct a series of over-flights over what has been dubbed "Ground Zero." Three remote sensors were used to collect data in need: black-and-white digital imagery at a resolution that can depict ground details down to 6 in. × 6 in., thermal information that can detect change in surface temperature of less than 1°C, and elevation data with a vertical accuracy of 6 in.

Figure 1.6 shows aerial images of the World Trade Center plaza and surrounding area before and after the attack of September 11.

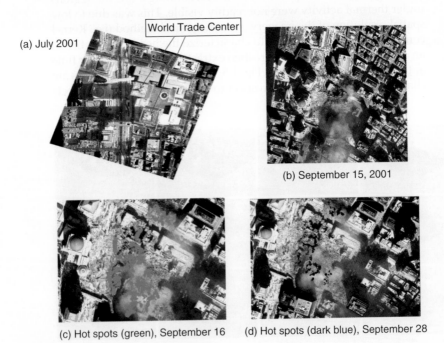

(a) July 2001

World Trade Center

(b) September 15, 2001

(c) Hot spots (green), September 16   (d) Hot spots (dark blue), September 28

FIGURE 1.6   Aerial photographs, in conjunction with thermal infrared data, were used in the recovery at the World Trade Center, New York City.

Figure 1.6(*a*) was taken in July 2000, and Fig. 1.6(*b*) was taken on September 15, 2001, with portions obscured by smoke. These images were collected using a digital camera system that can resolve objects 0.5 ft in diameter and larger. These images were then converted to map form (horizontally accurate to ±3 ft) through a computerized process of image projection called digital orthophotography. The resulting image shows all ground objects in their correct geographic position enabling planners and rescue workers to obtain precise measurements of distances between objects as well as accurate computations of area. These maps show current conditions at the site and provide planners with tools to direct equipment and personnel.

Figure 1.6(*c*) and (*d*) are computer composites of orthophotographs overlaid with an image captured using a thermal infrared camera system. The color composite overlays were generated using a thermal sensor that is sensitive to infrared radiation. They show the locations of hot spots within the debris field where there was a strong probability of lingering underground fires. A comparison of thermal images captured on September 16 and 28, 2001 indicates the hot spots had been reduced and became smaller, so it was now safer to approach the debris field.

1.6 Geographic Information Systems

1.6.1 Why Geographic Information Systems?

Geographic information system (GIS) has been used to enhance the functions of remote sensing image processing at various stages (Weng, 2009). It also provides a flexible environment for entering, analyzing, managing, and displaying digital data from various sources necessary for remote sensing applications. Many remote sensing projects need to develop a GIS database to store, organize, and display aerial and ground photographs, satellite images, and ancillary, reference, and field data.

GIS is computer software specifically designed for use with geographic data that performs a comprehensive range of data handling tasks, including data input, storage, retrieval, and output, in addition to a wide variety of descriptive and analytical functions (Calkins and Tomlinson, 1977). The backbone analytical function of GIS is overlay of spatially referenced data layers. Each layer describes a single geographical property within a bounded geographic area, such as soil, elevation, land use, trees, transportation network, rivers, and so on. The GIS overlay allows delineating the spatial relationships among these data layers.

GIS today is far broader and harder to define. Many people prefer to refer to its domain as Geographic Information Science & Technology (GIS&T), and it has become embedded in many academic and practical fields. The GIS&T field is a loose coalescence of groups of users, managers, academics, and professionals all working with geospatial information. Each group has a distinct educational and "cultural" background. Each identifies itself with particular ways of approaching particular sets of problems. Over the course of development, many disciplines have contributed to GIS. Therefore, GIS has many close and far "relatives." Disciplines that have traditionally researched geographic information technologies include cartography, remote sensing, geodesy, surveying, photogrammetry, etc. Disciplines that have traditionally researched digital technology and information include computer science in general and databases, computational geometry, image processing, pattern recognition, and information science in particular. Disciplines that have traditionally studied the nature of human understanding and its interactions with machines include cognitive psychology, environmental psychology, cognitive science, and artificial intelligence.

GIS has been called or defined as an enabling technology because of the breadth of uses in the following disciplines as a tool. Disciplines that have traditionally studied the Earth, particularly its surface and near surface in either physical or human aspect, include geology, geophysics, oceanography, agriculture, ecology, biogeography, environmental science, geography, global science, sociology, political science, epidemiology, anthropology, demography, and many more. When the focus is largely upon the utilization of GIS to attain solutions to real-world problems, GIS has more of an engineering flavor with attention being given to both the creation and the use of complex tools and techniques that embody the concepts of GIS. Among the management and decision-making groups, for instance, GIS finds its intensive and extensive applications in resource inventory and management, urban planning, land records for taxation and ownership control, facilities management, marketing and retail planning, vehicle routing and scheduling, etc. Each application area of GIS requires a special treatment, and must examine data sources, data models, analytical methods, problem-solving approaches, and planning and management issues.

1.6.2   Raster GIS and Capabilities

In GIS, data models provide rules to convert real geographical variation into discrete objects. Generally, there are two major types of data model: raster and vector. Figure 1.7 illustrates these two GIS data models. A raster model divides the entire study area into a regular grid of cells in specific sequence (similar to pixels in digital remote sensing imagery), with each cell containing a single value. The raster model is space filling because every location in the study area corresponds to a cell in the raster. A set of data describing a

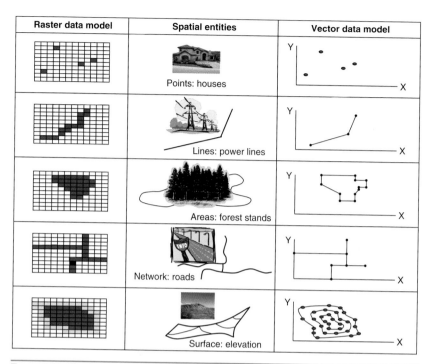

Raster data model	Spatial entities	Vector data model
	Points: houses	
	Lines: power lines	
	Areas: forest stands	
	Network: roads	
	Surface: elevation	

FIGURE 1.7   Raster and vector data models (*After Heywood et al. 1998*).

single characteristic for each location (cell) within a bounded geographic area forms a raster data layer. Within a raster layer, there may be numerous zones (also named patch, region, or polygon), with each zone being a set of contiguous locations that exhibit the same value. All individual zones that have the same characteristics form a class of a raster layer. Each cell is identified by an ordered pair of coordinates (row and column numbers), and does not have an explicit topological relationship with its neighboring cells.

Important raster datasets used in GIS include digital land use and land cover data, digital elevation models (DEMs) of various resolutions, digital orthoimages, and digital raster graphic (DRG) produced by the U.S. Geological Survey. Because remote sensing generates digital images in raster format, it is easier to interface with a raster GIS than any other type. It is believed that remote sensing images and information extracted from such images, along with GPS data, have become major data sources for raster GIS (Lillesand, Kiefer, and Chipman, 2008).

The analytical capabilities of a raster GIS result directly from the application of traditional statistics and algebra to process spatial data—raster data layers. Raster processing operations are commonly grouped into four classes, based on the set of cells in the input layers participating in the computation of a value for a cell in the output layer (Gao, Zhan, and Menon, 1996). The classes are: (1) per-cell or local operation, (2) per-neighborhood or focal operation, (3) per-zone or zonal operation, and (4) per-layer or global operation. A suite of raster processing operations may be organized logically to implement a particular data analysis application. Cartographic modeling is such a technique, which builds models within a raster GIS by using "map algebra" (Tomlin, 1991). The method of map algebra involves the application of a sequence of processing commands to a set of input map layers to generate a desired output layer. Many raster GIS of the current generation have made a rich set of functions available to the user for cartographic modeling. Among these functions are overlay, distance and connectivity mapping, neighborhood analysis, topographic shading analysis, watershed analysis, surface interpolation, regression analysis, clustering, classification, and visibility analysis (Berry, 1993; Gao et al., 1996).

1.6.3. Vector GIS and Capabilities

The traditional vector data model is based on vectors. Its fundamental primitive is a point. Lines are created by connecting points with straight lines and areas are defined by sets of lines. Encoding these features (point, lines, and polygons) is based on Cartesian coordinates using the principles of Euclidean geometry. The more advanced topological vector model, based on graph theory, encodes geographic features using nodes, arcs, and label points. The stretch of common boundary between two nodes forms an arc. Polygons are built by linking arcs, or are stored as a sequence of coordinates.

The creation of a vector GIS database mainly involves three stages: (1) input of spatial data; (2) input of attribute data; and (3) linking the spatial and attribute data. The spatial data are entered via digitized points and lines, scanned and vectorized lines, or directly from other digital sources. In the process of spatial data generation, once points are entered and geometric lines are created, topology must be built, which is followed by editing, edge matching, and other graphic spatial correction procedures. Topology is the mathematical relationship among the points, lines, and polygons in a vector data layer. Building topology involves calculating and encoding relationships between the points, lines, and polygons.

Attribute data are those data that describe the spatial features, which can be keyed in or imported from other digital databases. Attribute data are often stored and manipulated in entirely separate ways from the spatial data, with a linkage between the two types of data by a corresponding object ID. There are various database models used for spatial and attribute data in vector GIS. The most widely used one is the georelational database model, which uses a common feature identifier to link the topological data model (representing spatial data and their relationships) to a relational database management system, that is, DBMS (representing attribute data).

The analytical capabilities of vector GIS are not quite the same as with raster GIS. There are more operations dealing with objects, and measures such as area have to be calculated from coordinates of objects, instead of counting cells. The analytical functions can be grouped broadly into topological and non-topological ones. The former includes buffering, overlay, network analysis, and reclassification, whereas the latter includes attribute database query, address geocoding, area calculation, and statistical computation (Lo and Yeung, 2002).

1.7   Global Positioning System and Mobile Mapping

Global positioning system (GPS) is another essential technology when remote sensing projects need to have accurate ground control points and to collect *in situ* samples and observations in the field. In conjunction with GPS and wireless communicating technologies, mobile mapping becomes a technological frontier. It uses remote sensing images and pictures taken in the field to update GIS databases regularly in order to support problem solving and decision making at any time and any place.

A GPS receiver calculates its position by precisely timing the signals sent by the GPS satellites high above Earth. Each satellite continually transmits messages, which include the time the message was transmitted, precise orbital information (the ephemeris), and the general system health and rough orbits of all GPS satellites (the almanac). The receiver utilizes the messages it receives to determine the transit time of each message and computes the distances to each satellite. These distances along with the satellites' locations are used with the possible aid of trilateration to compute the position of the receiver (Fig. 1.8). Trilateration is a method of determining the relative positions of objects using the geometry of triangles. This position is then displayed, perhaps with a moving map display or latitude and longitude; elevation information may be included. Many GPS units also show derived information such as direction and speed, calculated from position changes.

Three satellites might seem enough to solve for position because space has three dimensions and a position on the Earth's surface can be assumed. However, even a very small clock error multiplied by the very large speed of light—the speed at which satellite signals propagate—can result in a large positional error. Therefore, receivers frequently use four or more satellites to solve for the receiver's location and time. The very accurately computed time is effectively hidden by most GPS applications, which use only the location. However, a few specialized GPS applications do use the

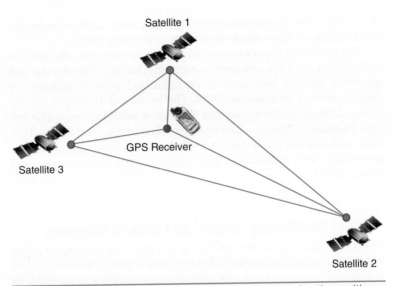

FIGURE 1.8   GPS uses the method of trilateration to determine the position of a receiver.

time; these include time transfer, traffic signal timing, and synchronization of cell phone base stations. Although four satellites are required for normal operation, fewer apply in special cases. If one variable is already known, a receiver can determine its position using only three satellites. For example, a ship or plane may have known elevation. Some GPS receivers may use additional clues or assumptions (such as reusing the last known altitude, dead reckoning, inertial navigation, or including information from the vehicle computer) to give a degraded position when fewer than four satellites are visible.

Two mobile mapping techniques are gaining momentum in the commercial domain and in the everyday life of the general public. One relates to the development of location-aware personal digital assistant (PDA), which consists of a GPS-equipped handheld computer or a palmtop GIS and may use such datasets as geographic features and attributes, aerial and ground photos, satellite images, and digital pictures (Gao, 2002). Another prospect for mobile mapping technology is distributed mobile GIS (DM-GIS) (Gao, 2002). In principle, the DM-GIS is very similar to PDA, and is typically composed of palmtop computers with GPS and a camera (Karimi, Krishnamurthy, Banerjee, and Chrysanthis, 2000). They communicate via a wireless network with a cluster of backend servers where GIS or remote sensing image data are stored. Digital pictures taken in the field can be relayed to the servers to update the GIS database as frequently as needed (Gao, 2002). In the future, a seamless integration of the four components, that is, remote sensing, GIS, GPS, and telecommunications would be possible, which allows for real-time imaging and processing in remote sensing and real-time GIS (Xue, Cracknell, and Guo, 2002). This real-time geospatial resolution is based on a real-time spatial database updated regularly by means of telecommunications to support problem solving and decision making at any time and any place. Real time in remote sensing imaging processing refers to the generation of images or other information to be made available for inspection simultaneously, or very nearly simultaneously, with their acquisitions (Sabins, 1997).

1.8   Google Earth

Google Earth is a virtual globe, map, and geographic information program that was originally called EarthViewer 3D, and was created by Keyhole, Inc., a company acquired by Google in 2004. It maps the Earth by the superimposition of satellite imagery, aerial photography, and GIS into a 3D globe. It allows users to see things like cities and houses looking perpendicularly down or at an oblique angle, that is, with a bird's eye view. The degree of resolution available is based on the points of interest and popularity, ranging from 15 m to 15 cm (6 in.). Google Earth allows users to search for addresses for some countries, to enter coordinates, or simply to use the mouse to browse to a location. For large parts of the Earth's surface, only 2D images are available from almost vertical photography. For other parts of the world, 3D images of terrain and buildings are available. Google Earth uses DEM data collected by NASA's Shuttle Radar Topography Mission. This means one can view the Grand Canyon or Mount Everest in three dimensions, instead of 2D like other areas. Google Earth supports managing 3D geospatial data through keyhole markup language (KML). Many urban buildings and structures from around the world now have detailed 3D view, including (but not limited to) those in the United States, Canada, Ireland, India, Japan, United Kingdom, Germany, and Pakistan. Recently, Google added a feature that allows users to monitor traffic speeds at loops located every 200 yards in real-time, and provide a street-level view in many locations. Most recently, images became available for the entire Earth's ocean floor. Google Earth 5 includes a separate globe of the planet Mars. The Martian terrain can be viewed and analyzed for research purposes. On July 20, 2009, the 40th anniversary of the Apollo 11 mission, Google introduced Google Moon, which allows users to view satellite images of the Moon.

Google Earth is available in 37 languages. Google Earth provides a unique tool for seeing our world and searching for geographic information. It is also useful for many civic applications such as checking roof conditions, getting information about a disaster area, or locating an inaccessible tropical jungle. However, it has also been criticized by a number of special interest groups, including national officials, as being an invasion of privacy and even posing a threat to national security. Some citizens have expressed concerns over aerial information depicting their properties and residences being disseminated freely.

1.9   Summary

Remote sensing, as the science and the technology of acquiring information about the Earth's surface and atmosphere using airborne and spaceborne sensors, has a long history. Contemporary remote sensing methods and techniques, however, were not developed until approximately four decades ago. Radar, hyperspectral, Lidar, and commercial remote sensing are rapidly expanding the frontiers of remote sensing from different perspectives. The applications of remote sensing are no longer limited to a few areas such as military, forestry, agriculture, and geology. Instead, its seamless integration with GIS, GPS, and mobile mapping allows remote sensing to apply to many diverse fields ranging from environment and resources, to homeland security, disasters, and everyday decision making, to name a few. Furthermore, Google Earth opens a new era for the public to use remote sensing data and to participate in collective sensing.

Key Concepts and Terms

Geographic information system (GIS) A computer system specifically designed for use with geospatial data that performs a comprehensive range of data handling tasks, including data input, storage, retrieval, and output, in addition to a wide variety of descriptive and analytical functions.

Global positioning system (GPS) GPS provides location and time information in all weather for any locality on Earth by computing its position through precisely timing the signals sent by the GPS satellites high above the Earth.

Lidar remote sensing Laser pulses are transmitted toward the ground and the distance between the sensor and the ground is measured based on the return time of each pulse.

Photogrammetry The science of obtaining reliable measurements through the medium of photography.

Photography   The art or process of producing images on a sensitized surface by the action of light or other radiant energy.

Radar remote sensing   Remotely sensed observations use the sun's energy in the microwave portion of the electromagnetic spectrum (approximately 1 mm to 1 m wavelength).

Remote sensing   Generally refers to the activities of recording/observing/ perceiving (sensing) objects or events without being in direct contact with the objects or events being observed.

Satellite remote sensing   The technology of acquiring information about the Earth's surface (land and ocean) and atmosphere using sensors onboard satellite platforms.

Thermal remote sensing   Remotely sensed observations using the energy in the infrared portion of the electromagnetic spectrum (approximately 3 to 14 µm wavelength).

Review Questions

1.   Remote sensing acquires images of the Earth's surface from above it. This is often referred to as a "bird's eye" view. What are some advantages of this view as opposed to a ground-level view (as you practice every day)?

2.   Identify one practical use of remote sensing technology, such as forestry, agriculture, hazards, and water resources, and explain how remote sensing can be applied to this particular area and list the major benefits.

3.   By reviewing the history of remote sensing, we learn that many sectors have contributed to its success today, from military, to government, industries, and commercial sectors. Can you speculate how future remote sensing technology may affect the everyday life of the public?

4.   Aerial photographs taken from airplanes usually have a clearer view of the target area than do images from satellite observations. For an application in which you are interested, such as agriculture, environment, geology, urban planning, or forestry, explain the advantages and disadvantages of each type of remote sensing technique.

5.   Why does remote sensing need GIS? How might the rapid development of remote sensing help GIS?

6.   Why is GPS called an enabling technology for remote sensing?

7.   Google Earth has largely changed the public view of remote sensing technology. Justify this statement by providing your reasons.

8.   Mobile mapping techniques are gaining momentum in the commercial domain and in the everyday life of the public. Provide some examples of the integration of remote sensing, GIS, GPS, and telecommunications technologies being crucial for problem solving.

References

Avery, T. E., and G. L. Berlin. 1992. *Fundamentals of Remote Sensing and Airphoto Interpretation*, 5th ed. Upper Saddle River, N.J.: Prentice Hall.

Bailey, G. B., D. T. Lauer, and D. M. Carneggie. 2001. International collaboration: the cornerstone of satellite land remote sensing in the 21st century. *Space Policy*, 17(3):161–169.

Berry, J. K. 1993. Cartographic modeling: the analytical capabilities of GIS. In Goodchild, M. F., B. O. Parks, and L. T. Steyaert, Eds. *Environment Modeling with GIS*. Oxford: Oxford University Press, pp. 58–74.

Calkins, H. W., and R. F. Tomlinson. 1977. *Geographic Information Systems: Methods and Equipment for Land Use Planning*. Ottawa: International Geographical Union, Commission of Geographical Data Sensing and Processing and U.S. Geological Survey.

Colwell, R. N. (Ed.). 1983. *Manual of Remote Sensing*, 2nd ed. Falls Church, Va.: American Society for Photogrammetry and Remote Sensing.

Devine, R. 1993. *The Sputnik Challenge*. New York: Oxford University Press.

Dwayne A. D., J. M. Logsdon, and B. Latell (Eds.). 1988. *Eye in the Sky: The Story of the Corona Spy Satellites*. Washington, D.C.: Smithsonian Books, pp. 143–156.

Gao, J. 2002. Integration of GPS with remote sensing and GIS: reality and prospect. *Photogrammetric Engineering and Remote Sensing*, 68(5):447–453.

Gao, P., C. Zhan, and S. Menon. 1996. An overview of cell-based modeling with GIS. In Goodchild, M. F., L. T. Steyaert, B. O. Parks, et al., Eds. *GIS and Environment Modeling: Progress and Research Issues*, Fort Collins, Colo.: GIS World, Inc., pp. 325–331.

Heywood, I., S. Cornelius, and S. Carver. 1998. *An Introduction to Geographical Information Systems*. Upper Saddle River, N.J.: Prentice Hall.

House, F. B., A. Gruber, G. E. Hunt, and A. T. Mecherikunnel. 1986. History of satellites missions and measurements of the Earth Radiation Budget (1957–1984). *Reviews of Geophysics*, 24:357–377.

Jensen, J. R. 2007. *Remote Sensing of the Environment: An Earth Resource Perspective*, 2nd ed. Upper Saddle River, N.J.: Prentice Hall.

Karimi, H. A., P. Krishnamurthy, S. Banerjee, and P. K. Chrysanthis. 2000. Distributed mobile GIS—challenges and architecture for integration of GIS, GPS, mobile computing and wireless communications. *Geomatics Info Magazine*, 14(9):80–83.

Kramer, H. J. 2002. Observation of the Earth and Its Environment: Survey of Missions and Sensors, 4th ed. Berlin: Springer Verlag.

Lillesand, T. M., R. W. Kiefer, and J. W. Chipman. 2008. *Remote Sensing and Image Interpretation*, 6th ed. New York: John Wiley & Sons.

Lo, C. P. and A. K. W. Yeung. 2002. *Concepts and Techniques of Geographic Information Systems*. Upper Saddle River, N.J.: Prentice Hall.

Melesse, A., Q. Weng, P. Thenkabail, and G. Senay. 2007. Remote sensing sensors and applications in environmental resources mapping and modeling. *Sensors*, 7:3209–3241.

Rajasekar, U. 2005. Image Mining within Meteosat Data: A Case of Modeling Forest Fire. Master Thesis, International Institute for Geo-Information Science and Earth Observation, Enschede, the Netherlands.

Sabins, F. F. 1997. *Remote Sensing: Principles and Interpretation*, 3rd ed. New York: W.H. Freeman and Company.

Stoney, W. 2005. A guide to the global explosion of land-imaging satellites: markets and opportunities? *Earth Imaging Source*, January/February, 2(1):10–14.

Tomlin, C. D. 1991. Cartographic modeling. In Maguire, D. J., M. F. Goodchild, and D. W. Rhind, Eds. *Geographical Information Systems: Principles and Applications*. London: Longman.

Weng, Q. 2009. *Remote Sensing and GIS Integration: Theories, Methods, and Applications*. New York: McGraw-Hill.

Wong, M. S., and J. E. Nichol. 2011. Remote sensing of aerosols from space: A review of aerosol retrieval using MODIS. In Weng, Q., Ed. *Advances in Environmental Remote Sensing: Sensors, Algorithms, and Applications*. Boca Raton, Fla.: CRC Press, pp. 431–448.

Xian, G. 2007. Mapping impervious surfaces using classification and regression tree algorithm. In Weng, Q., Ed. *Remote Sensing of Impervious Surfaces*, Boca Raton, Fla.: CRC Press, pp. 39–58.

Xue, Y., A. P. Cracknell, and H. D. Guo. 2002. Telegeoprocessing: the integration of remote sensing, Geographic Information System (GIS), Global Positioning System (GPS) and telecommunication. *International Journal of Remote Sensing*, 23(9):1851–1893.

CHAPTER 2

Electromagnetic Radiation Principles

2.1 Introduction

The main energy for remote sensing is solar energy, also known as electromagnetic radiation. Photons, the basic unit of electromagnetic radiation, not only move as particles but also as waves of different frequencies and wavelengths. When electromagnetic radiation travels to the Earth and strikes the Earth's surface, it is reflected, transmitted, or absorbed (Fig. 2.1). For any given material or surface, the amount of solar radiation that reflects, absorbs, or transmits varies with wavelength. This important property of matter makes it possible to identify and to separate different substances on the Earth's surface, and forms the basis for remote sensing technology. Before reaching a remote sensor, the electromagnetic radiation has to make at least one journey through the Earth's atmosphere, and two journeys in the case of active systems. Some of the electromagnetic radiation are scattered by atmospheric gases, aerosols, and clouds; while others are absorbed by carbon dioxide, water vapor, or ozone. This chapter first discusses basic properties of major portions of electromagnetic radiation, and then examines how solar radiation interacts with the features on the Earth's surface. This is followed by a detailed discussion of different effects that the atmosphere may generate on electromagnetic radiation and remote sensing images.

2.2 Principles of Electromagnetic Radiation

Electromagnetic radiation is a form of energy with the properties of a wave, with a major source from the sun. Solar energy traveling in the form of wavelengths at the speed of light (denoted as C, equal to 3×10^8 ms^{-1}) is known as the electromagnetic spectrum. The waves propagate through time and space in a manner rather like water waves, but oscillate in all directions perpendicular to their direction

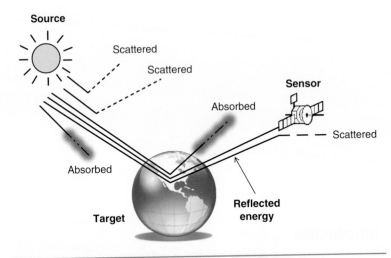

FIGURE **2.1**   Solar energy interaction with the environment.

of travel (Fig. 2.2). The fundamental unit of energy for an electromagnetic force is called a photon. The energy E of a photon is proportional to the wave frequency f as expressed in the following equation:

$$E = h f \qquad (2.1)$$

where the constant of proportionality h is Planck's Constant ($h = 6.626 \times 10^{-34}$ Joules-sec). Radiation from specific parts of the electromagnetic spectrum contains photons of different frequencies and wavelengths. Photons traveling at higher frequencies have more energy. Photons travel at the speed of light, that is, at 3×10^8 m per second. Photons not only move as particles but also move as waves,

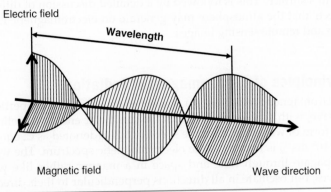

FIGURE **2.2**   Electromagnetic radiation travels in the form of a wave.

that is, they have a "dual" nature of particles and waves. When a photon travels as an electromagnetic wave, it has two components—one consisting of varying electric fields and the other of varying magnetic fields. Figure 2.2 illustrates this movement. The two components oscillate as sine waves mutually at right angles, and possess the same amplitudes (strengths), reaching their maxima-minima at the same time. Unlike other types of waves that require a carrier (e.g., water waves), photon waves can transmit through a vacuum (such as in space). When photons pass from one medium to another, for example, air to glass, their wave pathways are bent (follow new directions) and thus experience refraction (Short, 2009).

When any material on the Earth's surface interacts with incoming electromagnetic radiation, or is excited by internal processes, it will reflect, or emit, the amount of photons that differ at different wavelengths. Photon energy received by a remote sensor is commonly stated in power units such as Watts per square meter per wavelength unit. The plot of variation of power with wavelength gives rise to a specific pattern that is diagnostic of the material being sensed.

Electromagnetic waves are commonly characterized by two measures: wavelength and frequency. The wavelength (λ) is the distance between successive crests of the waves. The frequency (μ) is the number of oscillations completed per second. Other terms are also needed in order to better describe electromagnetic waves, including crest, trough, amplitude, and period (Fig. 2.3). Crest is the highest point of a wave, while trough is the lowest point. Amplitude is

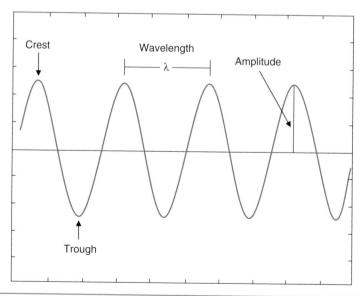

FIGURE 2.3   Key terms used for describing electromagnetic waves.

another important term in the wave description, referring to the distance from a wave's midpoint to its crest or trough. Period is the time it takes for two consecutive crests to pass a stationary point.

Wavelength and frequency are related by the following equation:

$$C = \lambda^* \mu \qquad (2.2)$$

Because the speed of light, C, is a constant, there exists an inverse relationship between the wavelength and the frequency of electromagnetic waves. As the wavelength becomes longer, the frequency decreases, and vice versa. In other words, a high-frequency electromagnetic wave (i.e., high-energy wave) has a shorter wavelength. Figure 2.4 further illustrates this inverse relationship between wavelength and frequency.

The electromagnetic spectrum, despite being thought of as a continuum of wavelengths and frequencies, may be divided into different portions by scientific conventions (Fig. 2.5). Major divisions of the electromagnetic spectrum, ranging from short wavelength, high frequency waves to long wavelength, low frequency waves, include gamma rays, X-rays, ultraviolet, visible, infrared spectrum, microwave, and radio wave. Therefore, in comparison of X-rays and microwave, X-rays have much shorter wavelength, higher frequency, and, therefore, higher energy.

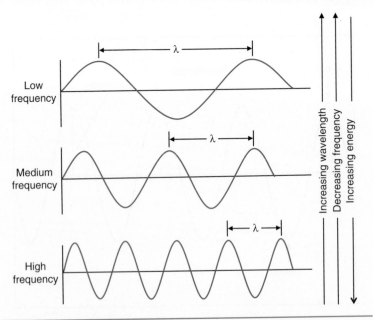

FIGURE 2.4   Inverse relationship between wavelength and frequency.

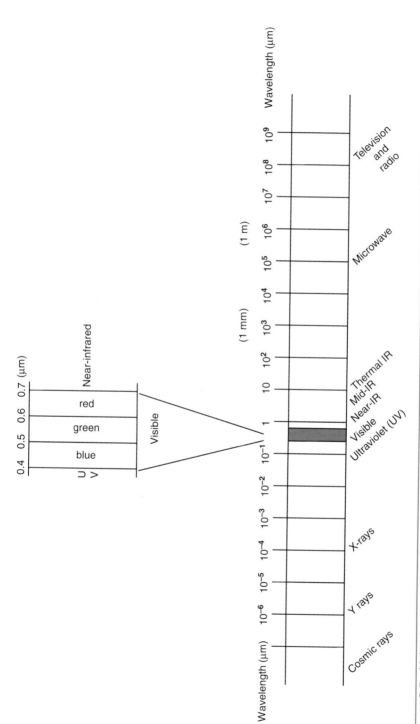

FIGURE 2.5 Major divisions of the electromagnetic radiation spectrum.

Among the major divisions of the electromagnetic spectrum, gamma rays have the shortest wavelength and highest frequency. Their wavelengths are less than approximately 10 trillionths of a meter, and are extremely penetrating. Gamma rays are generated by radioactive atoms and in nuclear explosions. These rays have been used in many medical, as well as astronomical applications. For example, images of our universe taken in gamma rays have yielded important information on the life and the death of stars and other violent processes in the universe. Gamma rays are not used in remote sensing because they are completely absorbed by the upper atmosphere in the process of traveling to the Earth (Sabins, 1997).

We are familiar with X-rays due to their wide use in the medical field. X-rays have wavelengths in the range from approximately 10 billionths of a meter to approximately 10 trillionths of a meter. Like gamma rays, X-rays are high-energy waves with great penetrating power. In addition to the medical field, they are also extensively utilized in weld inspections and astronomical observations. X-ray images of the sun provide important information on changes in solar flares that can affect space weather. X-rays from the sun are absorbed by the atmosphere before reaching the Earth, and therefore are not used in remote sensing.

Ultraviolet radiation has a range of wavelengths from 400 billionths of a meter (0.4 µm) to approximately 10 billionths of a meter (0.01 µm). Solar radiation contains ultraviolet waves, but most of them are absorbed by ozone in the Earth's upper atmosphere. A small dose of ultraviolet radiation that reaches the Earth's surface is beneficial to human health. Because larger doses of ultraviolet radiation can cause skin cancer and cataracts, scientists are concerned about the measurement of ozone and stability of the ozone layer. Only a small portion of ultraviolet radiation (in the range of 0.3 to 0.4 µm) can reach the Earth's surface, and can potentially be used for remote sensing of terrestrial features. However, this portion of ultraviolet radiation is easily scattered by the atmosphere owing to its short wavelength. This atmospheric effect reduces the usefulness of ultraviolet radiation in remote sensing of Earth materials (Campbell, 2007). Ultraviolet wavelengths are also used extensively in astronomical observatories.

The visible spectrum, commonly known as "sunlight," is the portion of the electromagnetic spectrum with wavelengths between 400 and 700 billionths of a meter (0.4 to 0.7 µm). This range is established by the sensitivity of human eyes to electromagnetic radiation. The majority of solar radiation reaches the top of the Earth's atmosphere as white light, which is separated into familiar rainbow colors when passing through a glass prism (Fig. 2.6). In remote sensing, however, the visible spectrum is usually divided into three segments: blue (0.4 to 0.5 µm), green (0.5 to 0.6 µm), and red (0.6 to 0.7 µm). The three segments refer to the additive primary colors, with which the

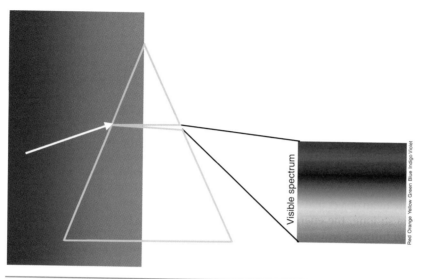

Figure 2.6   Visible spectrum when passing through a glass prism.

mixture of certain proportions of blue, green, and red can form all other colors. Approximately 40% of solar energy is concentrated in the visible spectrum, and it is relatively easy to go through the atmosphere reaching the Earth's surface (Avery and Berlin, 1992). Colors we see in nature are the result of reflection from particular objects. The maximum amount of electromagnetic radiation reflected from the Earth's surface is at the wavelength of 0.5 μm, corresponding to the green wavelength of the visible spectrum (Sabins, 1997). Therefore, the visible spectrum is very suitable for remote sensing of vegetation and for the identification of different objects by their visible colors in photography.

The infrared spectrum is the region of electromagnetic radiation that extends from the visible region to approximately 1 mm (in wavelength). Infrared waves can be further partitioned into near-infrared (0.7 to 1.3 μm), mid-infrared (1.3 to 3 μm), and far-infrared spectrum (3 to 1000 μm). However, due to atmospheric absorption and scattering, infrared radiation beyond 14 μm is not available for remote sensing. The infrared spectrum between 3 and 14 μm is also known as thermal radiation, or emitted radiation, and in plain language, "heat." In remote sensing, the near- and mid-infrared spectrums are conventionally combined to be called reflected infrared band, which acts similarly as the visible spectrum. The spectral property of reflected infrared radiation is very sensitive to vegetation vigor. Therefore, infrared images obtained by thermal sensors can yield important information on the health of crops and other types of vegetations. Solar energy contains a tiny amount of thermal infrared radiation. The thermal energy we use in remote sensing is mainly emitted by the Earth and the atmosphere. Thermal infrared radiation can help us monitor volcano activity, detect heat leaks from our houses, and see forest fires even when they are enveloped in an opaque curtain of smoke. Because the Earth emits thermal radiation day and night with the maximum energy radiating at the 9.7 μm wavelength (Sabins, 1997), some remote sensors have been designed to detect energy in the range from 8 to 12 μm in remote sensing of terrestrial features.

Microwave radiation has a wavelength ranging from approximately 1 mm to 1 m. Microwaves are emitted from Earth and the atmosphere, and from objects such as cars and planes. These microwaves can be detected to give information, such as on the temperature of an object that emits the microwave. Because their wavelengths are so long, the energy available is quite small compared to visible and infrared wavelengths. Therefore, the field of view must be large enough to detect sufficient energy in order to record a signal. Most passive microwave sensors are thus characterized by low spatial resolution. Active microwave sensing systems (such as radar) provide their own source of microwave radiation to illuminate the targets on the ground. A major advantage of radar is the capability of the radiation to penetrate through cloud cover and most weather conditions owing to its long wavelength. In addition, because radar is an active sensor, it can also be used to image the ground at any time during the day or night. These two primary advantages of radar, that is, all-weather and day or night imaging, make it a unique sensor.

Radio waves have wavelengths that range from approximately 30 cm to tens or even hundreds of meters. Radio waves can generally be utilized by antennas of appropriate size (according to the principle of resonance) for transmission of data via modulation. Television, mobile phones, wireless networking, and amateur radio all use radio waves. Radio waves can be made to carry data by varying a combination of the amplitude, frequency, and phase of the wave within a frequency band. The reflected radio waves can also be used to form an image of the ground features in complete darkness or through clouds.

2.3   Interactions with Earth Surface Materials

The electromagnetic radiation that impinges on the Earth's surface is called incident radiation. By interacting with features on the Earth's surface, the electromagnetic radiation is partitioned into three types, that is, reflection, transmission, and absorption. The proportion of the electromagnetic radiation that each type of interaction receives depends mainly on three factors: (1) the compositional, biophysical, and chemical properties of the surface feature/material; (2) the wavelength, or frequency, of the incident radiation; and (3) the angle at which the incident radiation strikes the surface (Avery and Berlin, 1992). Reflection is the bouncing of electromagnetic energy from a surface. Reflectance is the term to define the ratio of the amount of electromagnetic radiation reflected from a surface to the amount originally striking the surface. Transmission refers to the movement of energy through a surface. The amount of transmitted energy is wavelength dependent, and is measured as the ratio of transmitted radiation to the incident radiation known as transmittance. Some electromagnetic radiation is absorbed through electron or molecular reactions within the medium. Absorptance is the term to define the ratio between the amount of radiation absorbed by the medium (surface feature or material) and the incident radiation. According to the principle of energy conservation, we can establish the following two equations:

Incident radiation = reflected radiation + transmitted radiation
$$+ \text{absorbed radiation} \qquad (2.3)$$

$$\text{Reflectance} + \text{Transmittance} + \text{Absorptance} = 1 \qquad (2.4)$$

A portion of the absorbed energy is then re-emitted as emittance, usually at longer wavelengths, and the other portion remains and heats the target. Figure 2.7 illustrates the three types of interaction between electromagnetic radiation and the Earth's surface features using a leaf as an example.

When a surface is smooth, we get specular reflection where all (or almost all) of the energy is directed away from the surface in a single direction (Fig. 2.8). Specular reflectance within the visible wavelength ranges from as high as 0.99 for a very good mirror to as low as 0.02 to 0.04 for a very smooth water surface (Short, 2009).

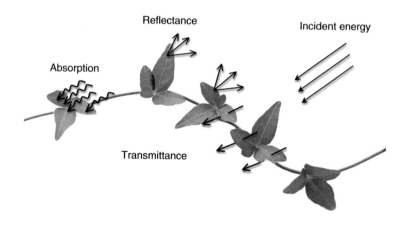

FIGURE 2.7   Three types of interaction between electromagnetic radiation and leaves.

FIGURE 2.8   Specular and diffuse reflection.

When the surface is rough and the energy is reflected almost uniformly in all directions, diffuse reflection occurs (Fig. 2.8). Most of the Earth's surface lies somewhere between a perfectly specular and a perfectly diffuse reflector. Whether a particular target reflects specularly or diffusely, or somewhere in-between, depends on the roughness of the surface in comparison to the wavelength of the incident radiation. If the wavelengths are much smaller than the surface variations or the particle sizes that make up the surface, diffuse reflection will dominate.

Remote sensing systems can detect and record both reflected and emitted energy from the Earth's surface by using different types of sensors. For any given material on the Earth's surface, the amount of solar radiation that reflects, absorbs, or transmits varies with wavelength. Theoretically speaking, no two materials would have the same amount of reflectance, absorption, and transmission within a specific range of wavelength. This important property of matter makes it possible to identify different substances or features and separate them by their "spectral signatures." When reflectance, absorption, or transmission is plotted against wavelength, a unique spectral curve is formed for a specific material. Figure 2.9 illustrates the typical spectral (reflectance) curves for three major terrestrial features—vegetation, water, and soil. Using their reflectance differences, we can distinguish these common Earth surface materials. When using more than two wavelengths, the plot in a multidimensional space can show

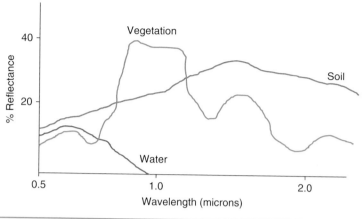

FIGURE 2.9   Typical spectral curves for three major terrestrial features—vegetation, water, and soil.

more separation among the materials. This improved ability to distinguish materials due to extra wavelengths is the basis for multispectral remote sensing.

2.4   Atmospheric Effects

Radiation detected by satellite remote sensors passes through the atmosphere with some path length. The atmosphere affects electromagnetic energy through the processes of absorption, scattering, and reflection. How much these processes affect the radiation along the path to the satellite depends on the path length, the presence of particulates and absorbing gases, and the wavelengths involved. Figure 2.10 illustrates how electromagnetic radiation is partitioned in the atmosphere.

Scattering is mainly caused by N_2, O_2 molecules, aerosols, fog particles, cloud droplets, and raindrops. Electromagnetic radiation is lost by being redirected out of the beam of radiation, but the wavelength does not change. Atmospheric scattering may impact remote sensing and resulted images in different ways, including causing skylight (allowing us to see in shadow), forcing images to record the brightness of the atmosphere in addition to the target on the Earth's surface, directing reflected light away from the sensor aperture or light normally outside the sensor's field of view toward the sensor's aperture (both decreasing the sharpness of the images), and making dark objects lighter and light objects darker (thus reducing the contrast).

Atmospheric scattering is accomplished through absorption and immediate re-emission of radiation by atoms or molecules. Rayleigh scattering occurs when radiation interacts with particles or molecules

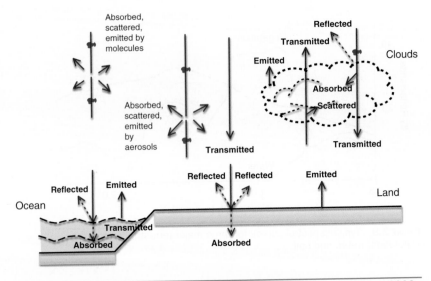

FIGURE 2.10   Process of atmospheric radiation (*After LaDue, J. and K. Pryor, 2003. NOAA/NESDIS*).

much smaller in diameter than the wavelength of the radiation. This type of scattering dominates in the upper 4 to 5 km of the atmosphere, but can take place up to 9 to 10 km above the Earth's surface. The degree of Rayleigh scattering is inversely proportional to the fourth power of the wavelength. As a result, short wavelength radiation tends to be scattered much more than long-wave radiation. The wavelength dependence and lack of directional dependence of Rayleigh scattering result in the appearance of "blue" sky.

If the atmospheric particulates are of roughly the same size as the wavelength of incident radiation, Mie scattering becomes important. This type of scattering takes place in the lower atmosphere (0 to 5 km) and can affect longer wavelength radiation. Smoke, dust, pollen, and water droplets are the dominant sources of Mie scattering. The orange and red light visible in the sky during sunrise and sunset are the result of this type of scattering. The greater the amount of smoke and dust particles in the atmosphere, the more that violet and blue light will be scattered away and only the longer orange and red wavelength light will be visible. Unlike Rayleigh scattering, most of the incident radiation is scattered in a forward direction because of Mie scattering. Mie scattering is the primary cause of haze, which often scatters sunlight into the sensor's field of view, degrading the quality of remote sensing images.

Non-selective scattering takes place in the lowest portions of the atmosphere where there are particles 10 times greater than the wavelength of the incident electromagnetic radiation. All wavelengths of light are scattered. The water droplets and ice crystals that comprise clouds scatter all visible light wavelengths equally well. Clouds appear white because of this type of scattering.

Atmospheric absorption is the process in which the electromagnetic radiation is absorbed and re-radiated again in all directions, and probably over a different range of wavelengths. Absorption can occur due to atmospheric gases, aerosols, clouds, and precipitation particles, but three types of atmospheric gasses mostly cause it—ozone, carbon dioxide, and water vapor. Ozone absorbs ultraviolet radiation, while carbon dioxide in the lower atmosphere absorbs energy in the spectral region of 13 to 17.5 µm. Water vapor in the lower atmosphere can be very effective in absorbing the energy in the spectrum between 5.5 and 7 µm and above 27 µm. The absorption by water vapor is an important consideration in remote sensing of humid regions. Overall, certain wavelengths of radiation are affected more by absorption than by scattering, especially in the infrared spectrum and wavelengths shorter than the visible portion of the electromagnetic spectrum. Where atmospheric absorption is high, the atmosphere must emit more radiation because of thermal excitation. Within these spectral regions, the surface of the Earth is not visible from space. However, these wavelengths are useful for observing (remote sensing) the atmosphere.

However, there are certain portions of the electromagnetic spectrum that can pass through the atmosphere with little or no attenuation (scattering and absorption). These spectral regions are called atmospheric windows. Four principal windows (by wavelength interval) open to effective remote sensing from above the atmosphere include: (1) visible-near infrared (0.4 to 2.5 µm); (2) mid-infrared (3 to 5 µm); (3) thermal infrared (8 to 14 µm); and (4) microwave (1 to 30 cm). Figure 2.11 shows these four atmospheric windows.

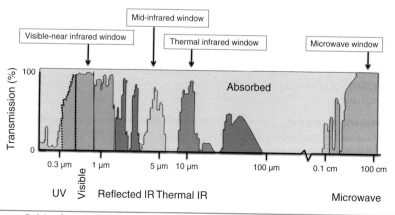

FIGURE 2.11   Atmospheric windows.

2.5 Summary

The electromagnetic spectrum may be divided into different portions based on wavelength and frequency, including gamma rays, X-rays, ultraviolet, visible, infrared spectrum, microwave, and radio wave. Electromagnetic radiation interacts with the Earth's surface features by reflection, absorption, or transmission. The amount of reflectance, absorption, and transmittance from a specific substance or material vary with wavelengths. The basic principle of remote sensing is recording the reflectance, in some case emittance because of absorption, by a remote sensor, and processing and interpreting acquired images in subsequent steps. Subsequent chapters of this book will describe the methods and techniques for processing and interpreting remote sensing images and data.

The Earth's atmosphere has major effects on remote sensing via scattering and absorption. In either case, the energy is attenuated in the original direction of the radiation's propagation. Therefore, the portions of the electromagnetic spectrum that can pass through the atmosphere with little or no attenuation are critical for remote sensing of Earth surface features. These atmospheric windows lie in the spectrum from visible to infrared and microwave.

The remote sensing principle described in this chapter is applied mainly to passive remote sensing systems, which record the reflected energy of the electromagnetic radiation or the emitted energy from the Earth by using cameras and thermal infrared detectors. Remote sensing can take another form via active sensing systems. The examples of active remote sensing include radar and Lidar. These remote sensing systems send out their own energy and record the reflected portion of that energy from the Earth's surface, and thus they can be used at daytime or nighttime and under almost all weather conditions. Chapter 10 focuses on active remote sensing.

Key Concepts and Terms

Absorptance   The ratio between the amount of radiation absorbed by the medium (surface feature or material) and the incident radiation.

Absorption   The portion of the electromagnetic radiation that is absorbed through electron or molecular reactions within the medium.

Atmospheric absorption   The process in which the electromagnetic radiation is absorbed by atmospheric gases, aerosols, clouds, and precipitation particles and re-radiated again in all directions, and probably over a different range of wavelengths.

Atmospheric scattering   Scattering is mainly caused by N_2, O_2 molecules, aerosols, fog particles, cloud droplets, and raindrops. Electromagnetic radiation is lost by being redirected out of the beam of radiation, but wavelength does not change.

Atmospheric window   The portion of the electromagnetic spectrum that transmits radiant energy effectively.

Diffuse reflection   Occurs when the surface is rough. The energy is reflected almost uniformly in all directions.

Electromagnetic radiation   A form of energy with the properties of a wave. The waves propagate through time and space in a manner rather like water waves, but oscillate in all directions perpendicular to their direction of travel.

Electromagnetic spectrum   The sun's energy travels in the form of wavelengths at the speed of light to reach to the Earth's surface. The energy is composed of several different wavelengths, and the full range of wavelengths is known as the electromagnetic spectrum.

Frequency   The number of oscillations completed per second.

Gamma rays   Wavelengths of gamma rays are less than approximately 10 trillionths of a meter. Gamma rays are generated by radioactive atoms and in nuclear explosions, and are used in many medical and astronomical applications.

Incident radiation   Electromagnetic radiation that interacts with an object on the Earth's surface.

Infrared spectrum   The region of the electromagnetic spectrum that extends from the visible region to approximately 1 mm in wavelength. Infrared radiation can be measured using electronic detectors, and has applications in medicine and in detecting heat leaks from houses, crop health, forest fires, and in many other areas.

Microwave radiation   Wavelengths range from approximately 1 mm to 1 m. Microwaves are emitted from the Earth, from objects such as cars and planes, and from the atmosphere. These microwaves can be detected to give information, such as the temperature of the object that emitted the microwaves.

Mie scattering   If the atmospheric particulates are of roughly the same size as the wavelength of incident radiation, Mie scattering becomes important. This type of scattering takes place in the lower atmosphere (0 to 5 km) and can affect longer wavelength radiation. Smoke, dust, pollen, and water droplets are the dominant sources of Mie scattering.

Non-selective scattering   Takes place in the lowest portions of the atmosphere where there are particles 10 times greater than the wavelength of the incident electromagnetic radiation. All wavelengths of light are scattered. The water droplets and ice crystals that comprise clouds scatter all visible light wavelengths equally well. Clouds appear white because of this type of scattering.

Photon   The fundamental unit of energy for an electromagnetic force.

Radio waves   Wavelengths range from less than 1 cm to tens or even hundreds of meters. Radio waves are used to transmit radio and television signals. The reflected waves can be used to create images of the ground in complete darkness or through clouds.

Rayleigh scattering   Occurs when radiation interacts with particles or molecules much smaller in diameter than the wavelength of the radiation. This type of scattering dominates in the upper 4 to 5 km of the atmosphere, but can take place in up to 9 to 10 km above the Earth's surface.

Reflectance   The proportional amount of the electromagnetic radiation reflected from a surface to the amount originally striking the surface.

Reflection   The bouncing of electromagnetic energy from a surface.

Spectral signatures   For any given material, the amount of solar radiation that reflects, absorbs, or transmits varies with wavelength. This important property of matter makes it possible to identify different substances or classes and separate them by their spectral signatures (spectral curves).

Specular reflection   Occurs when a surface is smooth. All (or almost all) of the energy is directed away from the surface in a single direction.

Transmission   The movement of light through a surface. Transmission is wavelength dependent.

Transmittance   The proportional amount of incident radiation passing through a surface. It is measured as the ratio of transmitted radiation to the incident radiation.

Ultraviolet radiation   The wavelength ranges from 400 billionths of a meter to approximately 10 billionths of a meter. Sunlight contains ultraviolet waves that can burn human's skin. Ultraviolet wavelengths are used extensively in astronomical observatories.

Visible spectrum   A portion of the electromagnetic spectrum with wavelengths between 400 and 700 billionths of a meter (400 to 700 nm). The rainbow of colors we see is the visible spectrum.

Wavelength   The distance between successive crests of waves.

X-rays   High-energy waves that have great penetrating power and are used extensively in medical and astronomical applications and in inspecting welds. The wavelength range is from approximately 10 billionths of a meter to approximately 10 trillionths of a meter.

Review Questions

1.   How does electromagnetic radiation interact with an object on the Earth's surface, for example, with a tree? Explain it by identifying different types of interactions.

2.   Human beings cannot "see" things in the dark, but some satellite sensors can. Explain why.

3. Among the three materials—vegetation, soil, and water—which materials have the highest and lowest reflectance in the visible spectrum (0.4 to 0.7 µm)? In which region of the spectrum—visible, near-infrared (0.7 to 1.3 µm), or mid-infrared (1.3 to 3 µm)—would the three materials be most easily separated?

4. In order to detect heat leaks from houses, or potential forest fires, which region of the spectrum should be used? Which regions of the spectrum have been used in astronomical observations?

5. What effect will atmospheric scattering have on satellite images? Please give an example of Rayleigh, Mie, and non-selective scattering.

6. What are atmospheric windows? How do they affect remote sensing?

7. The Amazon rain forest is the largest tropical jungle in the world. Clear cutting in this region has become a major environmental concern because it destructs habitats and disturbs the balance between the oxygen and carbon dioxide produced and used in photosynthesis. What would be some major difficulties for remote sensing of tropical forests there?

References

Avery, T. E. and G. L. Berlin. 1992. *Fundamentals of Remote Sensing and Airphoto Interpretation*, 5th ed. Upper Saddle River, N.J.: Prentice Hall.

Campell, J. B. 2007. *Introduction to Remote Sensing*, 4th ed. New York: The Guilford Press.

LaDue, J. and K. Pryor. 2003. Basics of Remote Sensing from Satellite. NOAA/ NESDIS Operational Products Development Branch, NOAA Science Center, Camp Springs, Md. http://www.star.nesdis.noaa.gov/smcd/opdb/tutorial/ intro.html (accessed May 27, 2010).

Sabins, F. F. 1997. *Remote Sensing Principles and Interpretation*, 3rd ed. New York: W.H. Freeman and Company.

Short, N. M. Sr. 2009. The Remote Sensing Tutorial, http://rst.gsfc.nasa.gov/ (accessed February 26, 2010).

CHAPTER 3

Characteristics of Remotely Sensed Data

3.1 Introduction

Regardless of passive or active remote sensing systems, all sensing systems detect and record energy "signals" from the Earth's surface features or from the atmosphere. Familiar examples of remote sensing systems include aerial cameras and video recorders. Sensing systems that are more complex include electronic scanners, linear/area arrays, laser scanning systems, etc. Data collected by these remote sensing systems can be either in analog format, for example, hardcopy aerial photography or video data, or in digital format, such as a matrix of "brightness values" corresponded to the average radiance measured within an image pixel. Digital remote sensing images may be directly input into a GIS for use. Analog data can also be used in GIS through an analog-to-digital conversion or by scanning. More often, remote sensing data are first interpreted and analyzed through various methods of information extraction to provide needed data layers for GIS. The success of data collection from remotely sensed imagery requires an understanding of four basic resolution characteristics, namely, spatial, spectral, radiometric, and temporal resolution (Jensen, 2005).

3.2 Data Recording Methods

3.2.1 Photographs

As discussed in Chap. 1, the output of a remote sensing system is usually an image representing the scene being observed. A remote sensing image is a 2D representation of part of the Earth's surface. Depending on the types of remote sensing devices used, remote sensing images may be analog or digital. Aerial photographs are examples of analog images, whereas satellite images typically belong to digital images.

Aerial photographs are made possible through a photographic process using chemical reactions on the surface of light-sensitive film to detect and record energy variations. Aerial photographs normally record over the wavelength range from 0.3 to 0.9 μm, that is, from ultraviolet to visible and near-infrared (IR) spectra. The range of wavelengths to be detected by a camera is determined by the spectral sensitivities of the film. Sometimes, filters are used in the camera in conjunction with different film types to restrict the wavelengths being recorded or to reduce the effect of atmospheric scattering. Mapping cameras used to obtain aerial photographs are usually mounted in the nose or underbelly of an aircraft. Aerial photographs can also be taken from space shuttles, unmanned aviation vehicles, balloons, or even kites. A mapping camera that records data photographically requires that the film be brought back to the ground for processing and printing. Figure 3.1 provides examples of black-and-white aerial photographs taken in the spring of 1998 (specifically the United States

(a) Airport   (b) Cropland

(c) Forest and lake   (d) Residential area

FIGURE 3.1   Examples of aerial photographs showing various land-use types.

Geological Survey Digital Orthophoto Quadrangle). These samplers illustrate different land use types in Indianapolis, Indiana, U.S.A.

3.2.2 Satellite Images

Satellite imagery may be viewed as an extension of aerial photography because satellite remote sensing relies on the same physical principles to acquire, interpret, and extract information content. Satellite images are taken at a higher altitude, which allows for covering a larger area of the Earth's surface at one time, and more importantly, satellite images are acquired by electronic scanners and linear/area arrays. Satellite sensors can usually detect and record a much wider range of electromagnetic radiation, from ultraviolet radiation to microwave. The electromagnetic energy is recorded electronically as an array of numbers in digital format, resulting in digital images. In a digital image, individual picture elements, called pixels, are arranged in a 2D array in columns and rows. Each pixel has an intensity value and a location address in the 2D array, representing the electromagnetic energy received at the particular location on the Earth's surface. Figure 3.2 illustrates pixels and how they make up digital images. The intensity value can have different ranges, depending upon the sensitivity of a sensor to the

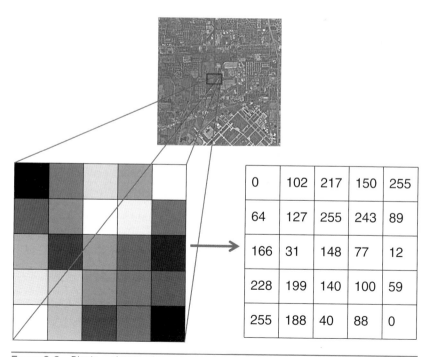

FIGURE 3.2   Pixels—picture elements made up of digital images.

changes in the magnitude of the electromagnetic energy. In most cases, a zero value indicates no radiation is received at a particular location and the maximum recordable number of the sensor indicates the maximum detectable radiation level. For example, if a satellite sensor has set the maximum detectable radiation level to be 255, it would have a range of 256 possible intensity levels in digital number. Then, when a computer displays a digital image acquired by this sensor, 256 different brightness levels (i.e., from 0 to 255) can possibly be used.

More than often, satellite sensors are designed to record the electromagnetic energy at multiple channels simultaneously. Each channel records a narrow wavelength range, sometimes known as a band. The digital image from each channel can be displayed as a black-and-white image in the computer, or we can combine and display three channels/bands of a digital image as a color image by using the three primary colors (blue, green, and red) of the computer. In that case, the digital data from each channel is represented as one of the primary colors according to the relative brightness of each pixel in that channel. Three primary colors combine in different proportions to produce a variety of colors as seen in the computer. Figure 3.3

FIGURE 3.3   A black-and-white satellite image of Beijing, China. This image was taken by SPOT, a French satellite sensor, on June 2, 2004.

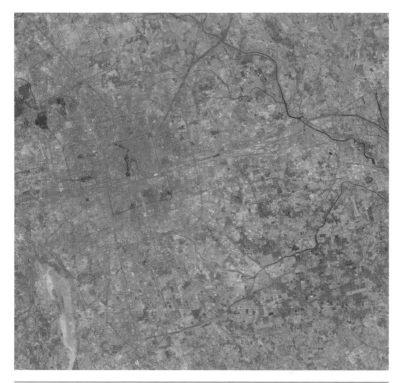

FIGURE 3.4   A false color composite of Beijing, which combines green, red, and near-IR bands of the image taken on June 2, 2004.

shows a black-and-white image of Beijing, the capital of China, and its surrounding areas. This image was taken by SPOT-5, a French satellite sensor, on June 2, 2004. Figure 3.4 shows a false color composite of Beijing, which combines green, red, and near-IR bands of the image taken on the same date of June 2, 2004. All bands have a spatial resolution of 10 m.

3.3   Spatial Resolution

Spatial resolution defines the level of spatial detail depicted in an image, and it is often defined as the size of the smallest possible feature that can be detected from an image. This definition implies that only objects larger than the spatial resolution of a sensor can be picked out from an image. However, a smaller feature may sometimes be detectable if its reflectance dominates within a particular resolution cell or it has a unique shape (e.g., linear features). Another meaning of spatial resolution is that a ground feature should be distinguishable as a separate entity in the image. However, the separation from neighbors or background is not always sufficient to identify the object. Therefore, spatial resolution involves both ideas of detectability and separability. For any feature to be resolvable in an image, it involves consideration of spatial resolution and spectral contrast, as well as feature shape.

Spatial resolution is a function of sensor altitude, detector size, focal size, and system configuration (Jensen, 2005). For aerial photography, spatial resolution is measured in resolvable line pairs per millimeter; while for other sensors, it refers to the dimensions (in meters) of the ground area that falls within the instantaneous field of view (IFOV) of a single detector within an array (Jensen, 2005). Figure 3.5 illustrates the concept of IFOV of a sensor as a measure of the ground area viewed by a single detector element in a given instant in time. The technical details of spatial resolution for aerial photographs are explored further in Chap. 4.

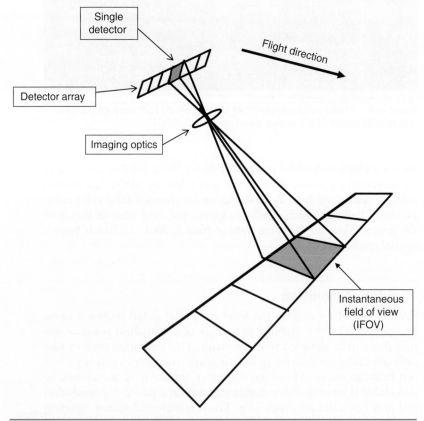

FIGURE 3.5   The concept of instantaneous field of view (IFOV).

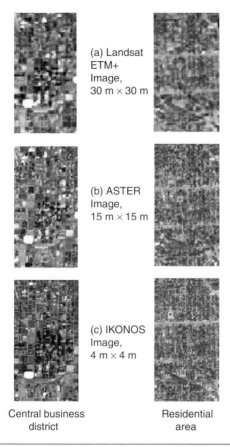

(a) Landsat
ETM+
Image,
30 m × 30 m

(b) ASTER
Image,
15 m × 15 m

(c) IKONOS
Image,
4 m × 4 m

Central business   Residential
district   area

FIGURE 3.6   Comparison of spatial resolution among Landsat ETM+, ASTER, and IKONOS images.

Spatial resolution determines the level of spatial details that can be observed on the Earth's surface. Coarse spatial-resolution images can detect only large features, while with fine resolution images, small features become visible. Figure 3.6 gives a comparison of a Landsat ETM+ image (spatial resolution = 30 m), an ASTER image (spatial resolution = 15 m), and an IKONOS image (spatial resolution = 4 m). These images were taken between 2000 and 2003 over the downtown area and a residential area in Indianapolis, Indiana. The IKONOS image gives fine details for each building in the downtown area and roads, houses, and vegetation in the residential area that Landsat and ASTER images cannot. However, for a remote sensing project, image spatial resolution is not the only factor needed to be considered. The relationship between the geographical scale of a study area and the spatial resolution of remote sensing image has to be studied (Quattrochi and Goodchild, 1997). At the local scale, high spatial-resolution imagery, such as IKONOS and QuickBird data, is more effective. At the regional scale, medium spatial-resolution data, such as Landsat TM/ETM+ and Terra ASTER data, are the most frequently used. At the continental or global scale, coarse spatial-resolution data, such as AHVRR and MODIS data, are most suitable. Higher resolution means the need for larger data storage and higher cost, and may introduce difficulties in image processing for a large study area. One goal for satellite sensors since the late 1990s has been improved spatial resolution, which now is down to better than 1 m. Sub-meter satellite images have introduced a new issue of residential privacy and have resulted in some legal cases.

As discussed previously, digital images are composed of pixels. It is important not to confuse the spatial resolution of a digital image and pixel size, although the two concepts are interrelated. In Fig. 3.7(a), a Landsat ETM+ image covering the city of Indianapolis, Indiana, United States, is displayed at full resolution. The spatial resolution of Landsat reflected bands is 30 m, and in Fig. 3.7(a), each pixel represents an area of 30 m × 30 m on the ground. In this case, the spatial resolution is the same as the pixel size of the image. In Fig. 3.7(b) to (f), the Landsat image is re-sampled to larger pixel sizes, ranging from 60 to 960 m. At the pixel size of 30 m, roads, rivers, and a central business district are visible. As the pixel size becomes larger, these features and objects become indiscernible. As discussed previously, pixel size is primarily defined by IFOV. This "optical pixel" is generally a circular or elliptical shape and combines radiation from all components within its ground area to produce the single pixel value in a given spectral channel (Harrison and Jupp, 2000). Along each image line, optical pixel values are sampled and recorded, and the average spacing of optical pixels relative to the ground area covered is then used to determine the geometric pixel size by using various mathematical algorithms (Harrison and Jupp, 2000). Coarse resolution images may include a large number of mixed pixels, where more than one surface cover type (several different features/objects) can be found within a pixel. Fine spatial resolution data considerably reduce the mixed pixel problem.

3.4   Spectral Resolution

In Chap. 2, we discussed the concept of spectral signature. Different features/materials on the Earth's surface interact with electromagnetic energy in different and often distinctive ways. A specific spectral response curve, or spectral signature, can be determined for each material type. Broad classes, such as vegetation, water, and soil, in an image can often be distinguished by comparing their responses over

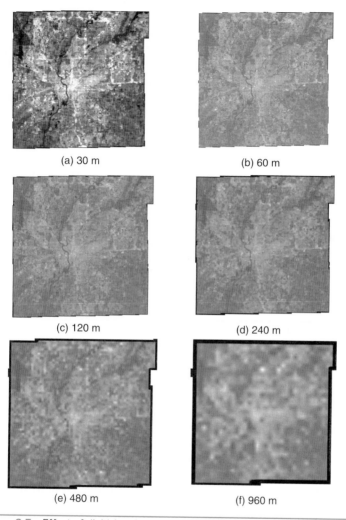

(a) 30 m   (b) 60 m

(c) 120 m   (d) 240 m

(e) 480 m   (f) 960 m

FIGURE 3.7   Effect of digitizing the same area with different pixel sizes.

distinct, very broad wavelength ranges. Spectral signature, or spectral response curve, is related to the spectral resolution of a remote sensor. Each remote sensor is unique with regard to what portion of the electromagnetic spectrum it detects and records. Moreover, each remote sensor records different numbers of segments of the electromagnetic spectrum, or bands, which may also have different bandwidths. Spectral resolution of a sensor, by definition, refers to the number and the size of the bands it is able to record (Jensen, 2005). For example, AVHRR, Onboard NOAA's Polar Orbiting Environmental Satellite (POES) platform, collects four or five broad spectral

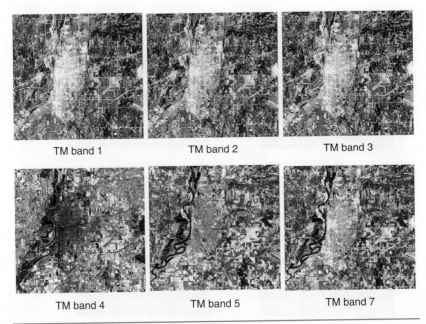

| TM band 1 | TM band 2 | TM band 3 |

| TM band 4 | TM band 5 | TM band 7 |

FIGURE 3.8   An example of Landsat TM spectral bands.

bands (depending upon individual instrument) in visible (0.58 to 0.68 μm, red), near-IR (0.725 to 1.1 μm), mid-IR (3.55 to 3.93 μm), and thermal IR portions (10.3 to 11.3 μm and 11.5 to 12.5 μm) of the electromagnetic spectrum. AVHRR, acquiring image data at the spatial resolution of 1.1 km at nadir, has been extensively used for meteorological studies, vegetation pattern analysis, and global modeling. Landsat Thematic Mapper (TM) sensor collects seven spectral bands. Its spectral resolution is higher than early instruments onboard Landsat such as MSS and RB). Figure 3.8 illustrates all reflective (excluding thermal IR band) bands of a Landsat TM image of Terre Haute, Indiana, U.S.A.

Most satellite sensors are engineered to use particular electromagnetic channels to differentiate between major cover types of interest, or materials, on the Earth's surface. The position, width, and number of spectral channels being sensed have been determined before a sensor is carried into the space. Table 3.1 shows the usages of each of the seven spectral bands of Landsat TM sensor. Spectral extent describes the range of wavelengths being sensed in all channels of an image. A remote sensing system may detect electromagnetic radiation in the visible and near IR regions, or extend to the middle and thermal IR regions. Microwave radiation may be recorded using radar and passive microwave sensors. An increase in

Name	Wavelength Range (mm)	Application
TM 1	0.45–0.52 (blue)	soil/vegetation discrimination; bathymetry/ coastal mapping; cultural/urban feature identification
TM 2	0.52–0.60 (green)	green vegetation mapping (measures reflectance peak); cultural/urban feature identification
TM 3	0.63–0.69 (red)	vegetated vs. non-vegetated and plant species discrimination (plant chlorophyll absorption); cultural/urban feature identification
TM 4	0.76–0.90 (near IR)	identification of plant/vegetation types, health, and biomass content; water body delineation; soil moisture
TM 5	1.55–1.75 (short IR)	sensitive to moisture in soil and vegetation; discriminating snow and cloud-covered areas
TM 6	10.4–12.5 (TIR)	vegetation stress and soil moisture discrimination related to thermal radiation; thermal mapping (urban, water)
TM 7	2.08–2.35 (short IR)	discrimination of mineral and rock types; sensitive to vegetation moisture content

TABLE 3.1   Landsat Thematic Mapper Bands and Usages

spectral resolution over a given spectral extent implies that a greater number of spectral channels are recorded. The finer the spectral resolution, the narrower the range of wavelength for a particular channel or band. The increased spectral resolution allows for a fine discrimination between different targets, but also increases costs in the sensing system, and data processing on board and on the ground. Therefore, the optimal spectral range and resolution for a particular cover type need to be modified with respect to practical uses of data collection and processing.

A major advance in remote sensor technology in recent years has been a significant improvement in spectral resolution, from bandwidths of tens of hundreds of nanometers (1 μm = 1000 nm), as pertains to the Landsat TM, to 10 nm or less. Hyperspectral sensors (imaging spectrometers) are the kind of instruments that acquire images in many very narrow, contiguous spectral bands throughout the visible, near-IR, mid-IR, and thermal-IR portions of the spectrum. Whereas Landsat TM obtains only one data point corresponding to the integrated response over a spectral band 0.24-μm wide, a hyperspectral sensor, for example, is capable of obtaining 24 data points over this range using bands approximately 0.01-μm wide.

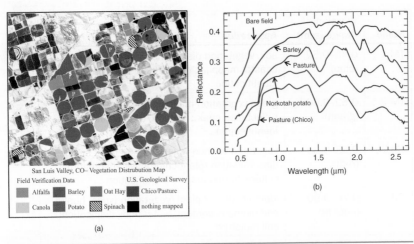

San Luis Valley, CO– Vegetation Distrubution Map
Field Verification Data   U.S. Geological Survey

▮ Alfalfa ▮ Barley ▮ Oat Hay ▮ Chico/Pasture

▮ Canola ▮ Potato ▨ Spinach ▮ nothing mapped

(a)

FIGURE 3.9   Vegetation species identification using hyperspectral imaging technique. (a) AVIRIS image showing circular fields in San Luis Valley, Colorado, and (b) reference spectra used in the mapping of vegetation species (*From Clark et al. 1995, with permission*).

NASA Airborne Visible-Infrared Imaging Spectrometer (AVIRIS) collects 224 contiguous bands with wavelengths from 400 to 2500 nm. A broadband system can only discriminate general differences among material types, while a hyperspectral sensor affords the potential for detailed identification of materials and better estimates of their abundance. Figure 3.9(a) shows an AVIRIS image of some circular fields in the San Juan Valley of Colorado. Vegetation or crop types in the fields, identified by different colors in the figure, are determined based on field survey and the spectral curves plotted in Fig. 3.9(b) for the crops indicated. Another example of a hyperspectral sensor is MODIS, on both NASA's Terra and Aqua missions and its follow-on to provide comprehensive data about land, ocean, and atmospheric processes simultaneously. MODIS has a two-day repeat global coverage with spatial resolution (250, 500, or 1000 m depending on wavelength) in 36 spectral bands.

3.5   Radiometric Resolution

Radiometric resolution refers to the sensitivity of a sensor to incoming radiance, that is, how much change in radiance there must be on the sensor before a change in recorded brightness value takes place (Jensen, 2005). Therefore, radiometric resolution describes a sensor's ability to discriminate subtle differences in the detected energy. Coarse radiometric resolution would record a scene using only a few brightness levels, that is, at very high contrast, whereas fine radiometric resolution would record the same scene using many brightness levels. Radiometric resolution in digital imagery is comparable to the number of tones in an aerial photograph (Harrison and Jupp, 2000). For both types of remote sensing images, radiometric resolution is related to the contrast in an image.

Radiometric resolution is frequently represented by the levels of quantization that is used to digitize the continuous intensity value of the electromagnetic energy recorded. Digital numbers in an image have a range from 0 to a selected power of 2 minus 1. This range corresponds to the number of bits (binary digits) used for coding numbers in computers. For example, Landsat-1 MSS initially records radiant energy in six bits (values ranging from 0 to 63, i.e., levels of quantization = 64 = 2^6), and was later expanded to seven bits (values ranging from 0 to 127, equal to 2^7). In contrast, Landsat TM data are recorded in eight bits, that is, its brightness levels range from 0 to 255 (equal to 2^8). The effect of changes in radiometric resolution on image feature contrast is illustrated in Fig. 3.10. Because human eyes can only perceive 20 to 30 different grey levels, the additional resolution levels provided by digital images are not visually discernible (Harrison and Jupp, 2000). Digital image processing techniques using computers are necessary in order to derive maximum discrimination from the available radiometric resolution.

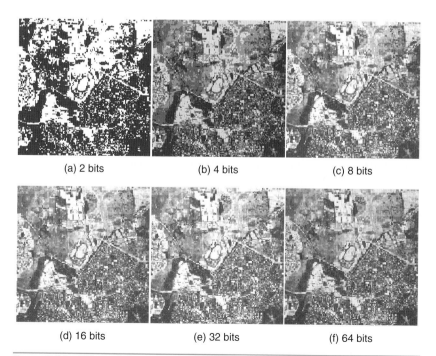

(a) 2 bits   (b) 4 bits   (c) 8 bits

(d) 16 bits   (e) 32 bits   (f) 64 bits

FIGURE 3.10   Effect of changes in radiometric resolution on image feature contrast (*From Harrison and Jupp, 2000*).

3.6   Temporal Resolution

Temporal resolution refers to the amount of time it takes for a sensor to return to a previously imaged location, commonly known as the repeat cycle or the time interval between acquisitions of two successive images. For satellite remote sensing, the length of time for a satellite to complete one entire orbital cycle defines its temporal resolution. Depending on the types of satellites, the revisit time may range from 15 min to more than 10 days. For air-borne remote sensing, temporal resolution is less pertinent because users can schedule flights for themselves.

Temporal resolution has an important implication in change detection and environmental monitoring. Many environmental phenomena constantly change over time, such as vegetation, weather, forest fire, volcano, and so on. These changes are reflected in the spectral characteristics of the Earth's surface, and are recorded in remote sensing images. Temporal differences between remotely sensed imagery are caused not only by the changes in spectral properties of the Earth's surface features/objects, but also they can result from atmospheric differences and changes in the sun's position during the course of a day and during the year. Temporal resolution is an important consideration in remote sensing of vegetation because vegetation grows according to daily, seasonal, and annual phenological cycles. It is crucial to obtain anniversary or near-anniversary images in change detection of vegetation. Anniversary images greatly minimize the effect of seasonal differences (Jensen, 2005). Many weather sensors have a high temporal resolution: GOES (every 15 min), NOAA AVHRR local area coverage (2 times per day), and Meteosat first generation (every 30 min). Observing short-lived, time-sensitive phenomena, such as floods, fire, and oil spills, requires truly high temporal resolution imagery. In general, ideal temporal resolution varies considerably for different applications. While monitoring of crop vigor and health may require daily images in a growing season, studies of urban development only need to have an image every 1 to 5 years (Jensen, 2007). Monitoring Earth's ecosystems, especially forests, is the major goal of many satellite missions. Human activities and natural forces can cause forest disturbance. Harvest, fire, and storm damage often result in abrupt changes, while changes due to insects and disease can last several years or longer. However, tree growth is always a slow, gradual process. Characterization of forest growth and disturbance therefore requires remote sensing data of different temporal resolutions (Huang et al., 2010).

The set overpass times of satellites may coincide with clouds or poor weather conditions. This is especially true in tropical areas, where persistent clouds and rains in the wet season offer limited clear views of the Earth's surface and thus prevent us from getting good quality images. Moreover, some projects require remote sensing images of particular seasons. In agriculture, phenology manifests itself through the local crop calendar—the seasonal cycle of plowing, planting, emergence, growth, maturity, harvest, and fallow (Campbell, 2007). The crop calendar stipulates us to acquire satellite images in the growth season, often from late spring to fall in the mid-latitude region, but the local climate needs to be considered. On the contrary, detection of urban buildings and roads may well be suited in the leave-off season in the temperate regions. In sum, the date of acquisition should be determined to ensure the extraction of maximum information content from remotely sensed data. When fieldwork needs to coincide with image acquisition, the date of acquisition must be planned well ahead of the time.

The off-nadir imaging capability of the SPOT-5 sensor reduces the usual revisit time of 26 days to 2 to 3 days depending on the latitude of the imaged areas. This feature is designed for taking stereoscopic images and for producing digital elevation models (Fig. 3.11). However, it obviously also allows for daily coverage of

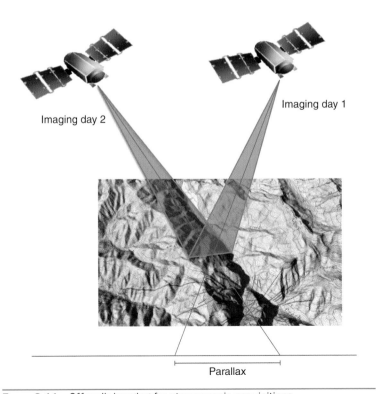

FIGURE 3.11   Off-nadir imaging for stereoscopic acquisitions.

selected regions for short periods, and provides another means for monitoring dynamic events such as flood or fire. The ASTER sensor has a similar capability of off-nadir imaging for its near infrared band (0.76 to 0.86 µm) by equipping a backward-looking telescope to acquire a stereo pair image.

3.7 Summary

To understand the characteristics of remotely sensed data, we must learn about the types of sensors used to retrieve these data and their functions and performance. In this chapter, we categorize remote sensors into two broad classes: analog and digital sensors. Aerial photographs, as an example of analog image, have the clear advantage of recording extremely fine spatial details. Satellite imagery, as an example of digital imagery, tends to have better quality in spectral, radiometric, and temporal resolution. The trend over the past 40 years has been directed toward improved resolution of each type. Considerable efforts have been made to design and construct sensors that maximize the resolution needed for their intended tasks. Superior spatial resolution permits ever-smaller targets, less than 1 m in size, to be seen as individuals. Greater spectral resolution means that individual entities (features, objects) can be identified more accurately because their spectral signatures are more distinguished as in hyperspectral sensors. Increased radiometric resolution offers sharper images, allowing for better separation of different target materials from their backgrounds. Finally, improved temporal resolution has made temporal analysis possible not only in the area of monitoring ecosystems and natural hazards, but also in detection of movement of ships and vehicles.

Under many situations, clear trade-offs exist between different forms of resolutions. For example, in traditional photographic emulsions, increases in spatial resolution are based on decreased size of film grain, which produces accompanying decreases in radiometric resolution, that is, the decreased sizes of grains in the emulsion portray a lower range of brightness values (Campbell, 2007). In multispectral scanning systems, an increase in spatial resolution requires a smaller IFOV, and thus less energy reaches the sensor. This effect may be compensated for by broadening the spectral window to pass more energy, that is, decrease spectral resolution, or by dividing the energy into fewer brightness levels, that is, decreasing radiometric resolution (Campbell, 2007). To overcome the conflict between spatial and temporal resolutions, NASA and Europe Space Agency have reportedly proposed to develop remote sensing systems that are able to provide data combining both high spatial resolution and revisit capabilities, among others.

Key Concepts and Terms

Aerial photographs   Remote sensing images in analog form made possible through the photographic process using chemical reactions on the surface of light-sensitive film to detect and record energy variations. Aerial photographs normally record over the wavelength range from 0.3 to 0.9 μm, that is, from ultraviolet to visible and near-IR spectra.

Image   Any pictorial representation, regardless of what wavelengths or remote sensing device has been used to detect and record the electromagnetic energy.

Image bands or channels   Information from a narrow wavelength range is gathered and stored in a channel, also sometimes referred to as a band.

Instantaneous field of view (IFOV)   A measure of the ground area viewed by a single detector element in a given instant in time.

Pixel   Picture element that has a single brightness value or digital number in an image, representing the electromagnetic energy received at the particular location on the Earth's surface. Pixel has both spatial and spectral property.

Radiometric resolution   The sensitivity of a sensor to incoming radiance, and describes a sensor's ability to discriminate subtle differences in the detected energy.

Spatial resolution   The size of the smallest object that can be resolved on the ground. In a digital image, the resolution is limited by the pixel size, that is, the smallest resolvable object cannot be smaller than the pixel size.

Spectral resolution   Each remote sensor is unique with regard to what portion of the electromagnetic spectrum it detects and records and the number of bands and their widths.

Temporal resolution   The amount of time it takes for a sensor to return to a previously imaged location, commonly known as the repeat cycle or the time interval between acquisitions of two successive images.

Review Questions

1. What are analog and digital images? How do aerial photographs differ from satellite images in respect to data recording method and the use of electromagnetic radiation energy?

2. Explain the relationship between pixels and images bands.

3. What is spatial, spectral, radiometric, and temporal resolution? Why do trade-offs exist between different forms of resolutions?

4. Why is it important to develop new sensors that improve more than one type of resolution?

5. Aerial photographs taken from airplanes usually have a clearer view of ground targets than digital images from satellite observations. For applications in agriculture and in military, respectively, explain the advantages and disadvantages of each type of remotely sensed data.

6. TM is a sensor of the Landsat program. How many spectral bands does it acquire and what are they? What is spatial resolution for its six reflected bands and that for the emitted band?

7. The pixel size of Landsat TM imagery is 30 m × 30 m, and the dimension of a complete scene is 185 km × 185 km. Calculate the number of pixels for a single band of a TM scene. Furthermore, calculate the number of pixels required to represent a single band of a Landsat MSS (79 m resolution) and make a comparison with Landsat TM imagery.

8. Which images have a better visual appearance, Landsat TM or IKONOS images? If you have a Landsat TM and an IKONOS image covering exactly the same area, which image would have a bigger file size when stored in computers?

9. Hyperspectral sensors can detect and record electromagnetic radiation in hundreds of very narrow spectral bands. What would be some of the advantages of these types of sensors? Any disadvantages?

10. A digital image has a radiometric resolution of 4 bits. What is the maximum value of the digital number that can be represented in that image? How about 6 and 12 bits?

11. Can you provide a couple of examples of using satellite sensors that have both high temporal-resolution (e.g., every 5 min) and high spatial-resolution (e.g., 1 m)?

References

Campell, J. B. 2007. *Introduction to Remote Sensing*, 4th ed. New York: The Guilford Press.

Clark, R. N., T. V. V. King, C. Ager, and G. A. Swayze. 1995. Initial vegetation species and senescence/stress mapping in the San Luis Valley, Colorado, using imaging spectrometer data. *Proceedings: Summitville Forum '95*, H. H. Posey, J. A. Pendelton, and D. Van Zyl, Eds., Colorado Geological Survey Special Publication 38, pp. 64–69. http://speclab.cr.usgs.gov/PAPERS.veg1/vegispc2.html.

Harrison, B. and D. Jupp. 2000. Introduction to remotely sensed data, a module of *The Science of Remote Sensing*, 4th ed., prepared by Committee on Earth Observation Satellites (CNES), the French Space Agency.

Huang, C., S. N. Goward, J. G. Masek, N. Thomas, Z. Zhu, and J. E. Vogelmann. 2010. An automated approach for reconstructing recent forest disturbance history using dense Landsat time series stacks. *Remote Sensing of Environment,* 114(1):183–198.

Jensen, J. R. 2005. *Introductory Digital Image Processing: A Remote Sensing Perspective,* 3rd ed. Upper Saddle River, N.J.: Prentice Hall.

Jensen, J. R. 2007. *Remote Sensing of the Environment: An Earth Resource Perspective,* 2nd ed. Upper Saddle River, N.J.: Prentice Hall.

Quattrochi, D. A., and M. F. Goodchild. 1997. *Scale in Remote Sensing and GIS.* New York: Lewis Publishers.

CHAPTER 4

Aerial Photo Interpretation

4.1 Introduction

Gaspard Felix took the first air photo in 1858. In 1909, airplanes were used as a platform for air photo cameras; only six years after the Wright brothers invented the airplane. Until spaceborne satellite remote sensing in the late 1950s, aerial photographs were a main source of information about the Earth's surface. Aerial photographs have the advantage of providing synoptic views of large areas. This characteristic also allows us to examine and interpret the Earth's surface features simultaneously over a large area and determine their spatial relationships, which is not possible from the ground. Aerial photographs are also a cost-effective tool in managing natural resources. Furthermore, aerial photographs have played a significant role in map-making, such as the United States Geological Survey's topographic maps.

Aerial photographs can be used to extract thematic and metric information, making it ready for input into GIS. Thematic information provides descriptive data about the Earth's surface features. Themes can be as diversified as their areas of interest, such as soil, vegetation, water depth, and land cover. Metric information includes location, height, and their derivatives, such as area, volume, slope angle, and so on. Thematic information can be obtained through visual interpretation of aerial photographs, while metric information can be extracted by using the principles of photogrammetry (refer to Chap. 5 for details).

4.2 Aerial Cameras

Aerial photos are obtained using mapping cameras that are usually mounted in the nose or the underbelly of an aircraft that then flies in discrete patterns or swathes across the area to be surveyed. Figure 4.1 illustrates the basic principle of a framing camera. A mapping camera is a framing system that captures an instantaneous "snapshot" of a

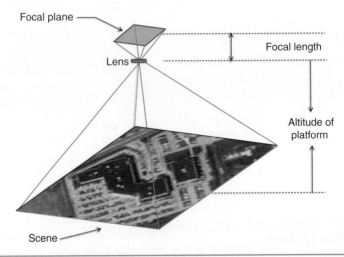

Focal plane

Focal length

Lens

Altitude of
platform

Scene

FIGURE 4.1   Basic principle of a framing camera.

scene on the Earth's surface. Camera systems are passive optical sensors that use a lens or a system of lenses, collectively referred to as the optics, to form a pictorial record or image at the focal plane. The aerial photo format (dimension) is associated with the size of the imaging area for a single frame.

Figure 4.2 illustrates a typical type of mapping camera. The major elements of a single-lens framing camera are shown in Fig. 4.3. These major elements include: (1) a camera body, that is, a light-proof chamber; (2) lens cone, which houses, from inside to outside,

FIGURE 4.2   A typical mapping camera (*Adapted from Short, 2009*).

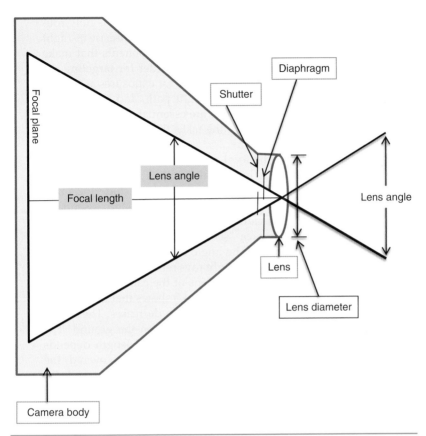

FIGURE 4.3   General structure of a single-lens framing camera (*After Avery and Berlin, 1992*).

the lens, shutter, diaphragm, and filter if any; (3) focal plane, to which the film is fixed and the light from a ground scene (i.e., the electromagnetic radiation) is directed and focused; and (4) film magazine, which contains an unexposed film reel and a take-up reel with main functions of holding the film in place, advancing the film between exposures, and winding-up the exposed film.

Aerial cameras are different from conventional cameras in order to have a more stable setup in the airplane (or space shuttle). This stability will ensure aerial photographs to have less geometric distortion, so that we can make reliable measurements from them. During the time the shutter is opened for exposure, the aerial camera is moving along with the airplane relative to the ground. This movement can cause image blur. Therefore, many aerial cameras include forward motion compensation (FMC) to minimize the effect of the movement. This equipment is especially important for high-speed airplanes (Avery and Berlin, 1992). Another component important to aerial mapping is a navigation control system. GPS is now routinely integrated into the camera system to provide a precise in-flight positional control. Additional fundamental components that make up an aerial camera system include a viewfinder for targeting the camera, an intervalometer that sets the rate for exposures as the airplane flies along the predetermined flight path, and an exposure control system. Of course, computers are essential nowadays in airplanes, which undertake many of the tasks that were once manual or mechanical.

The distance between the focal plane and the center of the lens when focused at infinity defines focal length (Fig. 4.4). The most common focal lengths used in aerial photography include 152.4 mm (6 in.), 210 mm (8¼ in.), and 305 mm (12 in.). Focal length essentially controls the angular field of view of the lens (similar to the concept of IFOV) and determines the area "seen" by the camera, that is, the ground coverage of a camera. The angle of ground coverage corresponds to the angle of the cone of light rays reflected from objects on the ground and passing through the lens of the camera before they expose the film in the focal plane. Figure 4.5 shows that the angle of ground coverage increases as the focal length decreases. The longer the focal length, the smaller the area covered on the ground. The choice for a camera of the most appropriate focal length depends on the type of application, the size of the area to be covered, the topography of the region under investigation, and the funds available.

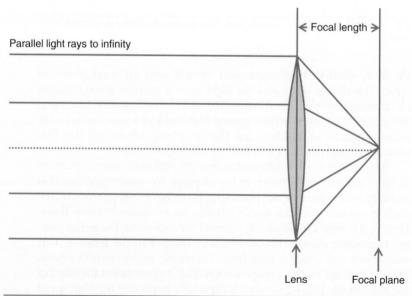

Figure 4.4   Focal length.

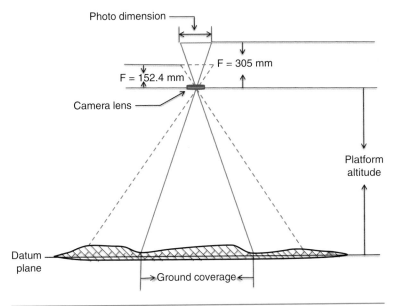

Photo dimension

F = 305 mm

F = 152.4 mm

Camera lens

Platform altitude

Datum plane

Ground coverage

FIGURE 4.5   Relationship between camera lens focal length and ground coverage.

Because wide-angle lenses (i.e., short focal length) exaggerate the displacement of tall objects, they are better suited for photographing flat terrain. Cameras with long focal length lenses are more proper for mapping those regions with sharp change in elevation. From Figs. 4.1 and 4.5, we can see that ground area coverage also depends on the altitude of the platform and the format of aerial photograph. At higher altitudes, a camera will be able to image a larger area on the ground, but with reduced details. The ratio of focal length to the altitude of the platform defines the scale of an aerial photograph, and has a close relationship with the spatial resolution of that photograph. We will discuss in detail the scale and spatial resolution of aerial photography in Section 4.5.

4.3   Films and Filters

Photographic film may be sensitive to wavelengths over a single range (such as black-and-white film) or in three wavelength ranges (such as three-layer color film). The four types of film that may be used for aerial photography are: (1) black-and-white visible, or "panchromatic"; (2) black-and-white near-IR; (3) color visible; and (4) false color near-IR (Harrison and Jupp, 2000). The spectral sensitivity of panchromatic film is in the range of 0.3 to 0.7 μm (visible and ultraviolet spectra), while black-and-white IR film can extend to

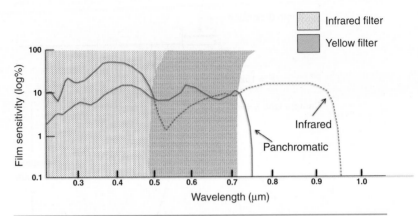

FIGURE 4.6   Spectral sensitivities of black and white visible ("panchromatic") and infrared films (*After Drury, 1987*).

approximately 0.9 μm in the near-IR region (Fig. 4.6). The IR film may be used with a dark red absorption filter to record near-IR radiation only or with appropriate filters to record selected regions in both visible and near-IR wavelengths (Harrison and Jupp, 2000). Figure 4.7 offers a good example of black-and-white near-IR aerial photograph, showing the Indianapolis Motor Speedway in Speedway, Indiana, United States, where the Indy 500, an American automobile race, is held annually over Memorial Day weekend. Each panel, (*a*) through (*d*), was taken to record the blue, green, red, and near-IR radiation.

Color films have been designed to mimic human vision; therefore, their three emulsion layers are designed to be sensitive to the three wavelength regions of blue, green, and red. Figure 4.8 exemplifies a natural color aerial photograph of the Indianapolis Motor

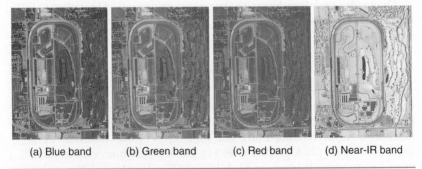

(a) Blue band   (b) Green band   (c) Red band   (d) Near-IR band

FIGURE 4.7   Example of black-and-white near-IR aerial photograph showing Indianapolis Motor Speedway. The aerial photograph was taken during the summer of 2008 via the National Agriculture Imagery Program. (*From Indiana Spatial Data Portal.*)

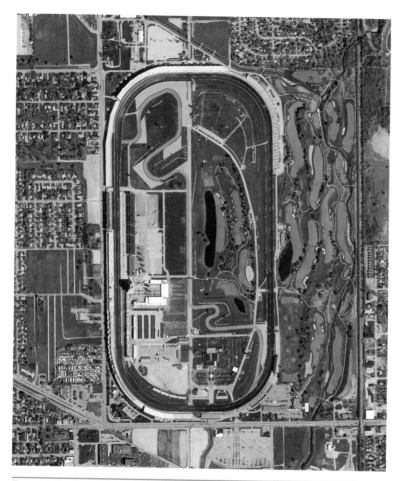

FIGURE 4.8   A natural color aerial photograph of the Indianapolis Motor Speedway, U.S.A.

Speedway. The spectral sensitivity of color IR films is different from normal color films (Fig. 4.9). Color IR films are used in conjunction with a yellow (i.e., blue absorbing) filter to record green, red, and IR radiation onto the three emulsion layers. Comparing with natural color films, the sensitivity of color IR films is extended to near-IR region (0.9 μm), and the broader range of shorter wavelengths is blocked by the yellow filter. Filters are used to block selected regions of wavelength, enabling color IR and natural color photography to capture specific reflectance characteristics of vegetation or other features of interest. In a false color photograph, colors are different from what we see in nature every day. Green vegetation appears red, deep clear water appears dark blue, and turbid water appears

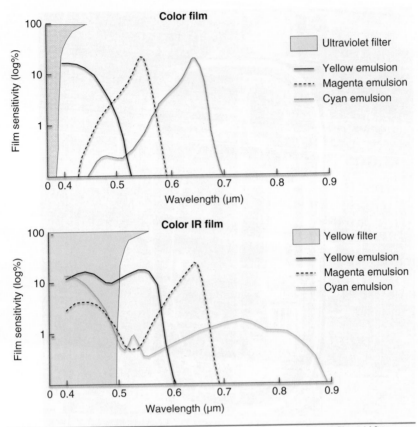

FIGURE 4.9   Spectral sensitivities of natural color and color infrared films (*After Drury, 1987*).

bright blue. Bare soil, roads, and buildings may appear in various shades of blue, yellow, or gray, depending on their composition. Figure 4.10 exemplifies a color IR aerial photograph of the Indianapolis Motor Speedway. False color images may be very useful in discriminating and interpreting certain features on the Earth's surface, such as vegetation.

Figure 4.11 illustrates a typical type of multispectral camera. Multispectral cameras use various combinations of film and filter to record different spectral regions in each aerial photograph. Continued experiments with the combination of different configurations of cameras, lenses, films, and filters help to select optimal wavelength regions, or "bands," for remote sensing of different features on the Earth's surface. The results of these experiments were later used in multispectral satellite remote sensing systems that will be discussed further in Chap. 7.

FIGURE 4.10   A color IR aerial photograph of the Indianapolis Motor Speedway, U.S.A.

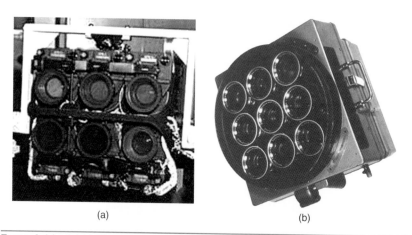

(a)   (b)

FIGURE 4.11   (a) Multispectral camera developed for use in the Skylab space station program. It uses separate lenses, each with its own narrow-band color filter, that are opened simultaneously to expose a part of the film inside the camera. (b) Nine-lens multispectral camera (Adapted from Short, 2009).

4.4   Methods of Aerial Photo Interpretation

Photographic interpretation refers to the act of examining aerial photographs or images for the purpose of identifying objects and judging their significance (Colwell, 1997). The activities of aerial photo/image interpreters may include: (1) detection/identification, (2) measurement, and (3) problem solving. In the process of detection and identification, the interpreter identifies objects, features, phenomena, and processes in the photograph, and conveys his or her response by labeling. These labels are often expressed in qualitative terms, for example, likely, possible, probable, or certain. The interpreter may also need to make quantitative measurements. Techniques used by the interpreter typically are not as precise as those employed by photogrammetrists. At the stage of problem solving, the interpreter identifies objects from a study of associated objects, or complexes of objects from an analysis of their component objects, and may examine the effect of some process to suggest a possible cause.

Seven elements are commonly used in photographic/image interpretation: (1) tone/color, (2) size, (3) shape, (4) texture, (5) pattern, (6) shadow, and (7) association. Tone/color is the most important element in photographic/image interpretation. Tone refers to each distinguishable variation from white to black, and is a record of light reflection from the land surface onto the film. The larger the amount of light received, the lighter the image on the photograph. Figure 4.12 shows a few examples of tones. Forest has a dark tone, houses and roads generally have bright tones, while croplands have various gray tones. Color refers to each distinguishable variation on an image produced by a multitude of combinations of hue, value, and chroma.

FIGURE 4.12   Example of various tones. Forest has dark tones, houses and roads have generally bright tones, while croplands have various gray tones. (*Image courtesy of U.S. Geological Survey.*)

Size provides another important clue in discrimination of objects and features. Both the relative and absolute sizes of objects are important. An interpreter should also judge the significance of objects and features by relating to their background. The shape of objects/features can provide diagnostic clues in identification. It is worthy to note that man-made features often have straight edges, while natural features tend not to.

Texture refers to the frequency of change and arrangement in tones. The visual impression of smoothness or roughness of an area can often be a valuable clue in image interpretation. For example, water bodies are typically fine textured, while grass is medium, and brush is rough, although there are always exceptions.

Pattern is defined as the spatial arrangement of objects. It is the regular arrangement of objects that can be diagnostic of features on the landscape. Man-made and natural patterns are often very different. Figure 4.13 shows an example of a typical low-density residential pattern that has been developed along a river. Residential communities typically have a mixture of houses, curvy roads, lawns, and water bodies, with the network of roads resembling a dendritic drainage pattern. Pattern can also be very important in geologic or geomorphologic analysis because it may reveal a great deal of information about the lithology and structural patterns in an area.

Shadow relates to the size and shape of an object. Geologists like low sun angle photography because shadow patterns can help identify objects. Steeples and smoke stacks can cast shadows that can facilitate interpretations. Tree identification can be aided by an examination of the shadows thrown. Association is one of the most helpful clues in

FIGURE 4.13   Example of pattern, showing a typical low-density residential area along a river. (*Image courtesy of U.S. Geological Survey.*)

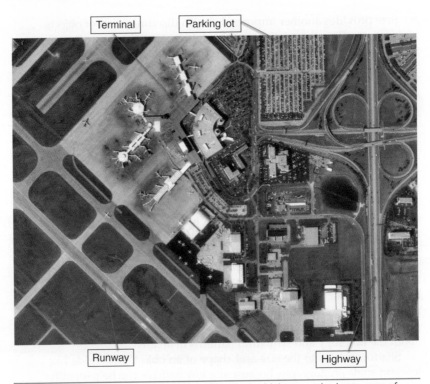

Terminal   Parking lot

Runway   Highway

FIGURE 4.14   Association of an airport with a major highway and a large area of parking space. (*Image courtesy of U.S. Geological Survey.*)

identifying man-made installations. Some objects are commonly associated with one another. Identification of one tends to indicate or confirm the existence of another. Smoke stacks, step buildings, cooling ponds, transformer yards, coal piles, and railroad tracks indicate the existence of a coal-fired power plant. Schools at different levels typically have characteristic playing fields, parking lots, and clusters of buildings in urban areas. Figure 4.14 shows association of an airport with a major highway and large area of parking space.

4.5   Scale of Aerial Photography

Scale is an important concept in understanding, interpreting, and analyzing aerial photographs, satellite images, and various types of maps. With knowledge of scale, it is possible to determine how much detail we can see from a remote sensing image, or a map, and how accurate our measurements can potentially be when estimating distance, height, area, and other metric information from such an image/map. Generally, scale can be defined as the relationship between map/image distance and actual ground distance in the Earth's surface. Scale may be represented in one of three methods: verbal scale, graphic/bar scale, and representative fraction (RF). Verbal scale is a descriptive statement, and it is known as "equivalent scale" (Paine and Kiser, 2003). For example, 1 inch on the photo/map equals 2000 feet on the ground, or 1 cm on the remote sensing image represents 1 km on the ground. Graphical/bar scale uses a line marked off in units of distance measured. This method is straightforward, and allows us to compare the straight-line distance between any two points on an image/map, by using a ruler, with the printed bar/graphic scale to determine the actual ground distance (Arnold, 2004). Graphical/bar scales are commonly seen in travel maps and various types of topographic maps produced by the United States Geological Survey. Representative fraction is a numerical statement of the scale relationship, and it is the ratio between the distance on the image/map to the distance on the ground. Representative fraction can be expressed as a simple fraction, for example, 1/24,000, or a ratio, for example, 1:24,000. The numerator should always be 1, and the denominator must have the same unit as the numerator. Therefore, the ratio is unitless.

There are different opinions in terms of which scales should be categorized as small, medium, or large. According to Paine and Kiser (2003), photo scales may be classified into four categories from the perspective of natural resources management, that is, small scale (1:24,000 and smaller), medium scale (1:10,000 to 1:24,000), large scale (1:1000 to 1:10,000), and vary large scale (1:1000 and larger). The U.S. Geological Survey topographic maps are often categorized into small scale (1:100,000,000 to 1:600,000), medium scale (1:600,000 to 1:75,000), and large scale (1:75,000 and larger). No matter which classification we use, large scale maps/images always show large amounts of detail over a small area, while small scale maps/images show small amounts of detail over a large area.

From Fig. 4.5, we learn that the scale of aerial photography may be determined by dividing the focal length of the camera by the vertical height of the lens above ground level (datum plane) based on the principle of similar triangles. The vertical height of the lens is also known as the flying height of an aircraft, that is, the platform altitude. Generally, a photograph is annotated with such information as the date and time of acquisition, the project code, the serial identification of the photograph (line number and exposure number), the focal length, and the flying altitude of the aircraft above mean sea level or above datum. When these parameters are present, the photo scale may be determined using the following formula:

$$PS = f/H \qquad (4.1)$$

where PS is the photo scale, f is the focal length of the camera used to take the photograph, and H is the flying altitude of the aircraft above the ground.

For example, a vertical photograph was taken at 4500 m above the ground. If the focal length of the camera used to take the photograph is 152.4 mm, the scale of the photograph is computed as follows:

$$PS = 152.4 \text{ mm}/4500 \text{ m}$$

$$= 152.4 \text{ mm}/(4500 \times 1000 \text{ mm})$$

$$= 1/29528 \text{ or } 1{:}29528$$

If focal length and flying height are unknown, the scale may also be computed by dividing the distance between two points on a photograph (PD) by the ground distance between the same two points on the ground (GD). Before computation takes place, the units used to measure these parameter pairs must be consistent. Photo scale may be expressed as:

$$PS = PD/GD \tag{4.2}$$

Similarly, map scale may be computed as:

$$MS = MD/GD \tag{4.3}$$

where PS is the photo scale, MS is the map scale, PD is the photo distance measured between two well-identified points on the photograph, MD is the map distance measured between two points on the map, and GD is the ground distance between the same two points on the photograph (or on the map), expressed in the same units.

For example, the distance between two points was measured to be 83.33 mm on a vertical photograph and 125.00 mm on a map. If the surveying ground distance between the same two points is 3000 m, the scales of the photograph and the map may be computed, respectively, as follows:

$$PS = 83.33 \text{ mm}/3000 \text{ m}$$

$$= 83.33 \text{ mm}/(3000 \times 1000 \text{ mm})$$

$$= 1/36000 \text{ or } 1{:}36000 \text{ and}$$

$$MS = 125 \text{ mm}/3000 \text{ m}$$

$$= 125 \text{ mm}/(3000 \times 1000 \text{ mm})$$

$$= 1/24000 \text{ or } 1{:}24000$$

4.6 Spatial Resolution of Aerial Photography

As we discussed in Chap. 3, spatial resolution involves both concepts of separability and detectability to determine the capacity of a film or lens to image spatial details on the ground. The spatial resolution of aerial photography is influenced by several factors, including mainly resolving power of a lens, resolving power of a film, ground target properties, atmospheric scattering, and the motion of aircraft during exposure (Sabins, 1997; Paine and Kiser, 2003). The resolving power of a film is largely determined by the size distribution of silver halide grains in the film emulsion. The smaller the grains are, the higher the resolution. Black-and-white films frequently have higher spatial resolution than do color films (Avery and Berlin, 1992). The resolving power of a lens is a function of the size and optical characteristics of the lens. Mapping cameras generally have better resolution than reconnaissance cameras (Avery and Berlin, 1992).

The resolving power of a film or lens is established in a controlled laboratory environment. The resolving power is measured using a standardized scheme that is based on "lines per millimeter." A typical tri-bar resolution target (Fig. 4.15) is photographed from a fixed altitude, and the number corresponding to the smallest number of line pairs visible (per millimeter) is recorded as its resolution. The camera system resolution, when considering any particular

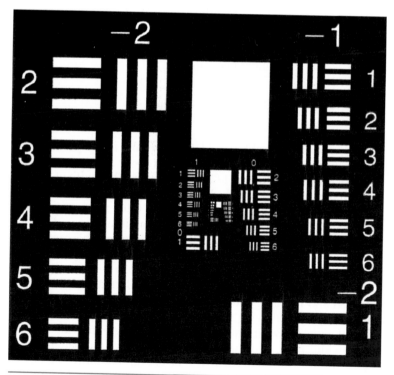

FIGURE 4.15   Typical tri-bar resolution chart of high contrast developed by the U.S. Department of the Army, Navy, and Air Force (1967).

lens-film combination, is set according to the lower-rated component. For example, if a film has the resolving power of 50 line pairs
per millimeter while the lens is capable of resolving 80 line pairs per
millimeter, the system resolution of the camera is then set by the
film.

If both focal length and flying height are known, we can use the
system resolution of a camera to determine ground resolution by
applying the following formula:

$$R_g = \frac{(R_s)(f)}{H} \tag{4.4}$$

where R_g is ground resolution in line pairs per meter, R_s is system
resolution in line pairs per millimeter, f is camera focal length in millimeter, and H is flying height in meter.

For example, if a camera with a 150-mm focal length takes aerial
photography at a flying height of 3000 m, and the system resolution
is 60 line pairs per millimeter, ground resolution is 3 line pairs per
meter:

$$R_g = \frac{\left(60 \text{ line}\frac{\text{pairs}}{\text{mm}}\right) \times (150 \text{ mm})}{3000 \text{ m}}$$

$$= \frac{9000 \text{ line pairs}}{3000 \text{ m}}$$

$$= 3 \text{ line pairs per meter.}$$

This means that the film could ideally resolve a ground resolution
pattern that is composed of 3 line pairs per meter. In other words, the
width of a single line pair is 1/3, or 0.33, meter. The width of a bar or
space in the line pair would be half of that value. That is, this camera
can resolve objects on the ground that are 0.165 m apart (approximately 6.5 in.), a value often referred to as minimum ground separation. This value also suggests that the smallest object this camera can
detect on the ground should have a dimension of 0.165 m.

The spatial resolution of aerial photography is also affected by
other factors, especially by the difference in reflected energy between
an object and its background. The spatial resolution is directly proportional to the following three factors (Paine and Kiser, 2003): the
contrast ratio between an object and its background, the object's
length-width ratio, and the number of objects in the pattern. If an
object has a similar tone/color with its background, it would be difficult to resolve. Therefore, the computation for ground resolution
using Eq. (4.4) often represents an approximation.

4.7   Summary

Cameras are the oldest sensors used for remote sensing of the Earth's surface. The use of aerial photography for interpreting and analyzing the Earth's surface features, patterns, and their dynamics, and for making reliable measurements has been a long tradition in remote sensing. To understand aerial photographs, we must have some fundamental knowledge about how they are made. In this chapter, we discuss the photographic process by introducing the basic structure of a mapping camera and its variants, films, and filters. The technological advances and countless experiments in camera, film, and filter over the past decades have led to modern digital satellite remote sensing. Two technological foci in aerial photography are the variety of film-filter combinations and camera system resolution. Both components are crucial in aerial photograph interpretation. To make scientific measurements from aerial photographs, we should also consider the height of platform for taking aerial photographs, the orientation of an aerial camera, tilt displacement during the flight, and topographic displacement. We have learned how to compute the scale of an aerial photograph, which closely relates to the system and ground resolutions. In the next chapter, we will pay closer attention to the spatial aspect of aerial photography—making reliable measurements.

Key Concepts and Terms

Angle of coverage   The angle of the cone of light rays reflected from objects on the ground and passing through the lens of the camera before they expose the film in the focal plane.

Color   Each distinguishable variation on an image produced by a multitude of combinations of hue, value, and chroma.

Color infrared photography   The three emulsion layers are sensitive to green, red, and the photographic portion of near-IR radiation, which are processed to appear as blue, green, and red, respectively. The colors in the color IR photographs give a "false" presentation of the targets relative to colors we see in the environment (e.g., trees appear red).

Focal length   The distance between the lens and the focal plane. Most common focal lengths are 152.4 mm (6 in.), 210 mm (8¼ in.), and 305 mm (12 in.).

Graphic or bar scale   The scale relationship on a map or an image is represented by a line or bar, which is divided into scaled units for the purpose of direct measurement on the map or the image, for example, feet, kilometers, and miles.

Natural color photography   Natural color photographs are designed to mimic human vision, and therefore their three emulsion layers are sensitive to the three wavelength regions of blue, green, and red. The colors in the natural color photographs resemble those we see in the environment (e.g., trees appear green).

Pattern   The spatial arrangement of objects, either man-made or natural. Patterns can be diagnostic of features on the landscape.

Photo interpretation   The act of examining aerial photographs or images for the purpose of identifying objects and judging their significance.

Representative fraction or ratio scale   The scale relationship on a map or an image is expressed as a fraction or ratio, for example, 1/24,000 or 1:24,000.

Scale   The relationship between the distance on a map or an image and actual ground distance of the same points.

Texture   The frequency of change and arrangement in tones.

Tone   Each distinguishable variation from white to black. A record of light reflection from the land surface on to the film. The more light received, the lighter the image is on the photograph.

Verbal scale or equivalent scale   A statement regarding the scale on a map or an image. For example, 1 in. on the map represents 300 feet on the ground.

Review Questions

1. What is the major function of a photographic filter?

2. In interpreting aerial photos, what are the seven elements (clues) often employed? Give an example in which we can use "shape" to differentiate natural features from man-made features. How can you easily differentiate roads from railroads in the aerial photographs?

3. Are color photographs more easily interpreted than black-and-white photographs? Explain the reasons.

4. Trees in an orchard often have a unique pattern. Do trees in the natural environment have a unique pattern, too? Compare the two patterns, and suggest the ways you would differentiate orchards from forests.

5. Give an example where you will find "association" to be an important clue in aerial photo interpretation.

6. Provide an example of usage for natural color photography and false color photography, respectively.

7. Explain the relationship between focal length and the angle of coverage. What type of focal length (6 in., 8.25 in., or 12 in.) would be most appropriate for aerial photography of a mountainous region?

8. Define map/image scale. Use an example to explain three types of scale representation on a map. Which map, 1:24,000 or 1:100,000, shows more details?

9. The distance, measured on a vertical photograph, between these two points is 8.65 cm. The same two points were identified on a 1:24,000 scale map and the distance was measured to be 2.73 cm. If the photograph was taken with a 15.24-cm focal length camera, determine: (*a*) the scale of the photograph; and (*b*) the flying height above the mean sea level at which the photograph was taken.

10. If an aerial photo was taken from an altitude of 5000 m with an aerial camera that has a 210-mm focal length, and if the system resolution is set to be 30 line pairs per millimeter, what would the ground resolution be?

11. What are major factors influencing the ground spatial resolution of aerial photography?

References

Arnold, R. H. 2004. *Interpretation of Airphotos and Remotely Sensed Imagery*. Long Grove, Ill.: Waveland Press.

Avery, T. E., and G. L. Berlin. 1992. *Fundamentals of Remote Sensing and Airphoto Interpretation*, 5th ed. Upper Saddle River, N.J.: Prentice Hall.

Colwell, R. N. 1997. History and place of photographic interpretation. In Philipson, W. R. *Manual of Photographic Interpretation*, 2nd ed. Bethesda, Md.: American Association of Photogrammetry and Remote Sensing.

Drury, S. A. 1987. *Image Interpretation in Geology*. London: Allen & Unwin Ltd.

Harrison, B., and D. Jupp. 2000. Introduction to remotely sensed data, a module of *The Science of Remote Sensing*, 4th ed., prepared by Committee on Earth Observation Satellites (CNES), the French Space Agency.

Paine, D. P., and J. D. Kiser. 2003. *Aerial Photography and Image Interpretation*. Hoboken, N.J.: John Wiley & Sons.

Sabins, F. F. 1997. *Remote Sensing Principles and Interpretation*, 3rd ed. New York: W.H. Freeman and Company.

Short, N. M., Sr. 2009. *The Remote Sensing Tutorial*, http://rst.gsfc.nasa.gov/ (accessed February 26, 2010).

CHAPTER 5

Photogrammetry Basics

5.1   Introduction

Aerial photographs have the advantage of providing us with a synoptic view of large areas. This characteristic makes it possible to examine and interpret ground objects simultaneously on a large area and to determine their spatial relationships. Aerial photographs are also cost effective in interpreting and managing natural resources, when comparing with field surveys. Historically, aerial photographs have played a significant role in map making and data analysis. All of these values are related to the discipline of photogrammetry.

Photogrammetry is traditionally defined as the science or art of obtaining reliable measurements by means of photography (Colwell, 1997). Recent advances in computer and imaging technologies have transformed traditional analog photogrammetry into digital (softcopy) photogrammetry, which uses modern technologies to produce accurate topographic maps, orthophotographs, and orthoimages by using the principles of photogrammetry. Photogrammetry for topographic mapping is normally applied to a stereopair of vertical aerial photographs (Wolf and Dewitt, 2000). To make geometrically corrected topographic maps out of aerial photographs, relief displacement must be removed by using the theory of stereoscopic parallax. Another type of error in an aerial photograph is caused by tilts of the aircraft around the x, y, and z axes of the aircraft (so-called roll, pitch, and yaw, see Fig. 5.1) at the time the photograph is taken (Lo and Yeung, 2002). All the errors with photographs can now be corrected using a suite of computer programs.

5.2   Types of Aerial Photos

From space, an aerial camera can be positioned to image the ground from different angles. Depending on the orientation of the camera relative to the ground during the flight, most aerial photographs are

93

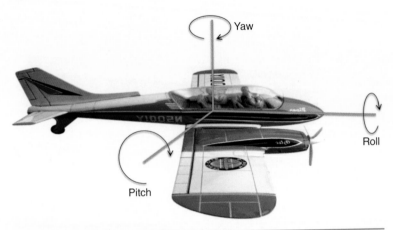

Figure 5.1   Three types of tilts of an aircraft around its x, y, and z axes—roll, pitch, and yaw.

classified as either vertical or oblique photographs. While vertical aerial photographs are acquired with the camera pointed vertically down (or slightly tilted, less than 3° from the vertical) to the Earth's surface, oblique aerial photographs are taken with the camera pointed to the side of the aircraft (Fig. 5.2).

Vertical photography is the most common use of aerial photography for remote sensing purposes. The ground area imaged by vertical photography is square in shape (Fig. 5.2a). Vertical photographs contain limited geometric distortions, and are good for accurate measurements and interpretation tasks. Figure 5.3 provides an example of vertical photography. When two consecutive vertical photographs in a flight line are overlapped, the overlapping area can be viewed stereoscopically to generate a 3D view of the terrain. This chapter focuses on the discussion of vertical photography.

The ground area imaged by oblique photography is trapezoidal in shape (Fig. 5.2b and 5.2c). Oblique aerial photographs can be divided further into two types—low oblique and high oblique. High-oblique photographs usually include the ground surface, horizon, and a portion of sky, while low-oblique photographs do not show horizon. The view of ground area is similar to what we see when looking through the window of an airplane. Distance features in oblique photographs look smaller than near features. Figures 5.4 and 5.5 illustrate a low-oblique and a high-oblique photograph, respectively. Oblique photographs are often taken for imaging a very large area in a single photograph and for depicting terrain relief (i.e., topography). They are not frequently used for measuring and mapping as ground features are distorted in scale from the foreground to the background.

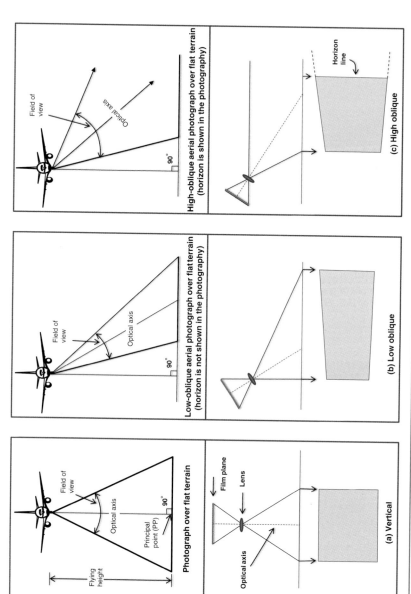

FIGURE 5.2 Orientation of an aerial camera for vertical, low- and high-oblique photography, and the geometry of imaged ground area for each type of aerial photography (*After Avery and Berlin, 1992*).

Figure 5.3 Example of a vertical photograph showing part of Plainfield, Indiana, U.S.A. (*Courtesy of U.S. Geological Survey.*)

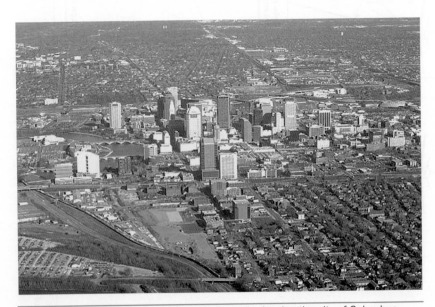

Figure 5.4 Example of a low-oblique photograph showing the city of Columbus, Ohio, U.S.A. (*Courtesy of Ohio Department of Transportation.*)

FIGURE 5.5   Example of a high-oblique photograph taken from the International Space Station in early April 2003. Mount Kilimanjaro is a dormant stratovolcano and the highest mountain in Tanzania, Africa. Kibo Summit (5893 m) at the top of Kilimanjaro is one of the few peaks in Africa to retain glaciers. This photograph provides a 3D perspective on the positions of the glaciers on the upper northwestern and southern flanks of the mountain. (*Courtesy of NASA's Earth Observatory.*)

5.3   Geometry of Aerial Photography

As discussed in Chap. 4, incoming light rays from objects on the ground pass through the camera lens before they are imaged on the film in the focal plane. Focal length is the term used to describe the distance between the focal plane and the center of the lens. Focal length modulates the ground angular coverage, and is a key element in defining the scale of an aerial photograph. The distance between the lens and the ground surface is referred to as flying height/ altitude. Figure 5.6 shows the basic geometry of a vertical aerial photograph. The x coordinate axis is arbitrarily assigned to the imaginary flight line direction on the photograph and the y-axis is assigned to a line that is perpendicular to the x-axis. These two axes usually correspond to the lines connecting the opposite fiducial marks recorded on each side of the photograph (i.e., positive image). The photo coordinate origin, O, is located in the intersection of the two lines joining the fiducial marks. Therefore, the point O coincides exactly with the geometric center of the photograph, which is called the principal point (PP). On any real aerial photography, there is no mark of a PP, but we can locate the PP by connecting fiducial marks at the opposite sides,

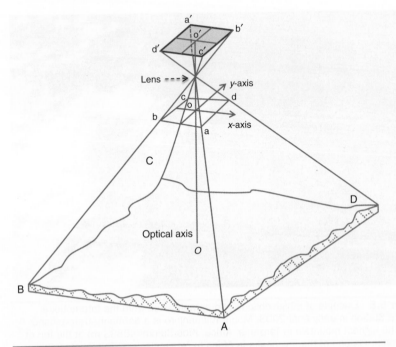

<small>FIGURE</small> **5.6**   Geometry of a vertical aerial photograph.

or by connecting opposite corners of the aerial photograph. These lines will all intersect at the PP (Fig. 5.7).

The principal point can be transferred stereoscopically onto the adjacent (left and right) photographs of the same flight line by

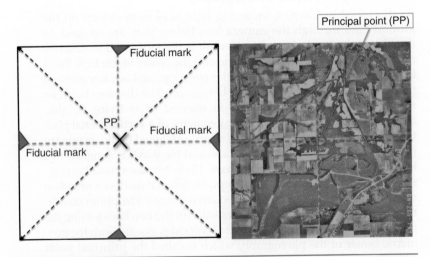

<small>FIGURE</small> **5.7**   Determination of principal point by connecting fiducial marks, or by connecting opposite corners of an aerial photograph.

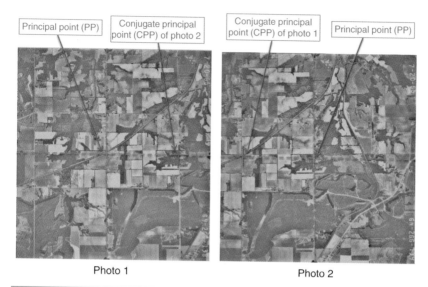

| Principal point (PP) | Conjugate principal point (CPP) of photo 2 | Conjugate principal point (CPP) of photo 1 | Principal point (PP) |

Photo 1   Photo 2

FIGURE 5.8 Determination of conjugate principal points (CPP) by transferring the principal points stereoscopically onto the adjacent (left and right) photographs of the same flight line by pricking through its transferred positions.

pricking through its transferred positions. These transferred points are called conjugate principal points (CPP). Figure 5.8 shows the method for locating CPP. The line segment joining the principal points and the conjugate principal points constitute the flight line of the aircraft, also called base line or air base. The determination of air base is important as it is used for lining up a photograph to adjacent photographs on the same flight line. The alignment of adjacent aerial photos is the first step to see them correctly in stereoscopy, to determine height and difference in the elevation of objects, and to make photogrammetric measurements. These measurements allow for making topographic maps. Because of distortions and image displacement (discussed later in this chapter), the distance between the PP and the CPP of the adjacent photograph will often be different.

The nadir, also called vertical point or plumb point, is the intersecting point between the plumb line directly beneath the camera center at the time of exposure and the ground (Fig. 5.9). Nadir is important in determining the height of ground objects in photographs based on the principle of parallax. However, it is not easy to locate the nadir in an aerial photograph. In a vertical aerial photograph, we often assume that the PP and nadir coincide. On a tilted aerial photograph, the nadir and the principal point have different positions, and the nadir is always on the down side of the tilted photograph.

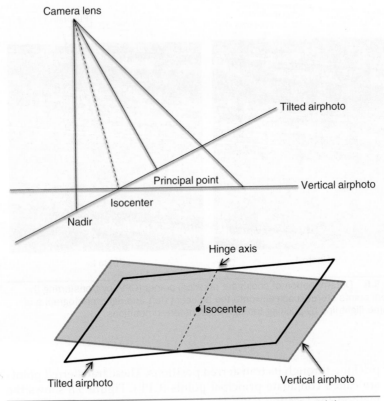

FIGURE 5.9   Locations of the principal point (PP), the nadir (n), and the isocenter (i) on a tilted vertical aerial photograph (*After Paine and Kiser, 2003; Campbell, 2007*).

Locating the nadir on tilted photographs usually requires sophisticated stereoscopic plotting techniques involving expensive instruments and ground control information (Fahsi, 1996).

If the nadir and the principal point have been located and if the direction of the flight is known, then we can determine the direction of tilt of the aircraft (Fig. 5.9). The isocenter is the point halfway between the principal point and the nadir and on the line segment joining these two points on the photograph. It is a point intersected by the bisector of the angle between the plumb line and the optical axis. The intersection of the plane of the tilted photograph and the plane of the desired vertical photograph forms a line that lies at right angles to the line connecting the principal point and the nadir. The isocenter is located at the intersection of these two lines (Fig. 5.9). The isocenter is the point from which tilt displacement is radial. On a true vertical photograph, the PP, the nadir, and the isocenter all coincide at the geometric center (i.e., PP) of the photograph. In this case, there is no tilt displacement, but relief displacement in the aerial photograph.

5.4   Image Displacement and Height Measurement

A vertical photograph is not a map owing to the optical characteristics inherent in a vertical aerial photograph. Maps use an orthogonal projection while aerial photographs use central projection (Fig. 5.10). On a planimetric map, all features on the ground surface are considered to be portrayed at their correct horizontal positions. The features are viewed from directly overhead, and thus have a truly vertical view of every detail. The spatial relationship among the features is accurate (Fahsi, 1996). In contrast, the perspective viewing of the camera results in a perspective or central projection on the photograph. Ground features are viewed as though they are all from the same point. Therefore, the images of most ground objects on vertical aerial photographs are generally displaced from their true plane positions. This type of displacement is known as image displacement in aerial photography.

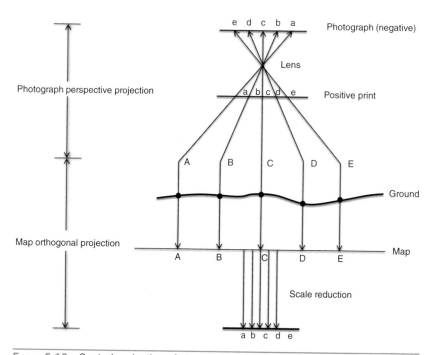

FIGURE 5.10   Central projection of photographs versus orthogonal projection of maps (*After Fahsi, 1996*).

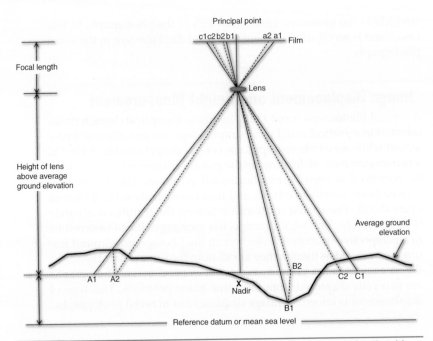

FIGURE 5.11   Effect of relief displacement on a vertical photograph (*After Arnold, 2004*).

The most pronounced image displacement is associated with topography or relief, that is, the difference in the relative elevations of ground objects. Relief displacement can affect most objects on the ground, including mountains, gorges, tall buildings, and even trees, with the exception of calm water surfaces (Avery and Berlin, 1992). Figure 5.11 shows the effect of topographic relief on image displacement on a vertical photograph. Ground features above a specified ground datum plane are displaced on the aerial photographs away from the PP, while those below the datum plane are displaced toward the PP. Relief displacement is radial from this point. In other words, the farther an object is away from the nadir, the larger amount the relief displacement will be. Relief displacement is a function of the distance of the displaced image from the nadir, the height of the object, and the scale of the aerial photograph. This relationship can be expressed in mathematical terms to determine the heights of ground objects. In addition, we can also utilize image displacement to view overlapping photographs in three dimensions (discussed in the next section).

5.4.1   Height Determination from a Single Aerial Photo

The heights of tall objects, if not pictured near the center of an aerial photograph, can be measured using only one photo. The method of height determination from a single aerial photo assumes (Avery and Berlin, 1992):

1. The principle point coincides with the photo nadir (assuming a *true* vertical photo).
2. The flying height/altitude from which the photo was acquired can be determined.
3. Both the top and the bottom of the object to be measured are clearly visible.
4. The degree of image displacement must be great enough to be measured accurately with available equipment (e.g., an engineering scale, or a precise ruler).

If these conditions are met, the height of an object can be computed by using the following equation:

$$h = \frac{d}{r(H)} \tag{5.1}$$

where h = height of the object; H = flying altitude above the surface where the photo is taken; d = length of the displaced image; and r = radial distance from the photo nadir to the top of the object.

For example, if a building on a "perfectly" vertical photograph has a distance between its top and its bottom of 0.5 in., the distance from the photo center to the top of the building is measured to be 3 in., and if the flying height of the aircraft is 1800 feet above the base of the building, the height of the building can be computed as follows:

$$h = (0.5 \text{ in.}/3 \text{ in.}) \times 1800 \text{ feet}$$

$$= 300 \text{ feet}$$

It should be noted that the measurements of d and r must be in the same unit, and H is expressed in the unit desired for the height of the object.

5.4.2 Height Determination Using Stereoscopic Parallax

In case it is difficult or we are unable to measure the length of a displaced image and/or the distance between the top and bottom of the object due to the quality of an aerial photograph, we can use another method to measure the heights of objects from aerial photographs. This method requires two overlapping air photos in the same flight line and utilizes stereoscopic parallax. Figure 5.12 shows an example of stereo pairs of aerial photographs for downtown Champaign, Illinois, United States. The equation for computing

FIGURE 5.12   Example of a stereopair of aerial photographs, showing downtown Champaign, Illinois, U.S.A. (*Prepared by the University of Illinois Committee on Aerial Photography, 1969.*)

the height of an object using the stereoscopic parallax method is as follows:

$$h = \frac{(H)dP}{P + dP} \tag{5.2}$$

where h = the height of the object of interest; H = flying altitude where the photo is taken above datum; dP = differential parallax; and P = absolute parallax.

Therefore, if the altitude of the aircraft above datum is known or can be calculated and if we can calculate the differential and the absolute parallax from the available stereo pairs, then we can determine the heights of objects in the photos. Parallax refers to the apparent displacement of the position of an object, caused by a shift in the point of observation. Absolute parallax is the sum of the differences between the base of conjugate objects and their respective nadirs and is always measured parallel to the air base (Avery and Berlin, 1992). Differential parallax is defined as the difference in the absolute stereoscopic parallax at the top and the base of the object on a pair of photographs. To measure parallaxes, we need to have a parallax-measuring device (e.g., a parallax bar), a stereoscope, and an engineer's scale. Figure 5.13 illustrates a stereoscope and a parallax bar. For vertical air photos, we can substitute the average photo-base length of the stereo pairs for P (absolute parallax) in the solution of Eq. (5.2).

Figure 5.14 shows a stereogram of the Washington Monument, Washington, D.C. We can measure the height of the Washington Monument from this stereo pair. The photo scale at the base of the monument is 1:4600. The camera focal length was 12 in. Therefore,

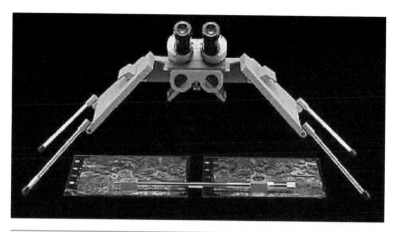

FIGURE 5.13 A stereoscope and a parallax bar used to measure parallaxes. (*Short, 2009.*)

the flying height was 4600 feet (12 in. divided by the scale 1/4600). The average photobase (P) of the stereopair is calculated to be 4.40 in. Absolute stereo parallax at the base and at the top of the monument is measured parallel to the line of flight with an engineer's scale. The difference is (2.06 in. −1.46 in.). This gives a dP of 0.60 in. Therefore,

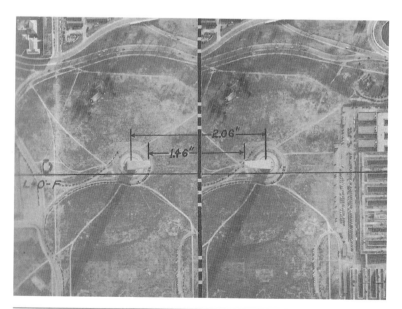

FIGURE 5.14 A stereogram of the Washington Monument, Washington, D.C. Scale: 1:4600 at the base of the monument (*Avery and Berlin, 1992*). (*Use with permission.*)

0.60 in. is the differential parallax of the displaced images. Substituting these values into Eq. (5.2):

$$h = (H)\frac{dP}{P+dP} = (4600 \text{ feet})\frac{0.60 \text{ inch}}{4.40 - 0.60 \text{ inch}} = 552 \text{ feet}$$

The actual height of the Washington Monument is 555 feet. Therefore, this is a very accurate measurement for this type of exercise. It should be noted that the flying height of the aircraft should be in the same unit as the object height. Differential parallax and absolute parallax, dP and P, should also be in the same units.

5.5   Acquisition of Aerial Photos

As discussed in Chap. 1, aerial photography represents the earliest technology in remote sensing. There are many existing aerial photographs in the United States and other countries, some traced back to the 1930s. Libraries, universities, and research labs may have developed their own achieves for historical photographs in analog, digital, or both forms, which may be requested to be reproduce for use with minimal cost. Therefore, any projects may consider using existing photos, but researchers need to note that some of them may be of poor quality, with the wrong scale, or on undesired film or print type. Some federal government agencies, such as the U.S. Geological Survey (USGS), U.S. Department of Agriculture (USDA), National Archives and Records Administration (NARA), and most states operate repositories of historical aerial photographs. The USGS has developed an Internet-based data search and acquisition tool called "Earth Explorer" (http://edcsns17.cr.usgs.gov/EarthExplorer/). Worldwide remotely sensed data can be searched based on geographical location (latitude and longitude), but some data may be limited to U.S. users only. For each type of aerial photograph, the USGS Earth Resources Observation and Science Center (http://eros.usgs.gov/#/Find_Data/Products_and_Data_Available/Aerial_Products) has a detailed production description along with the status of digitization, search, and download methods. The USDA's Aerial Photography Field Office has created a consolidated depository of over 10 million images for all aerial photographs that were obtained by the Agricultural Stabilization and Conservation Service, the U.S. Forest Service, and the Soil Conservation Service from 1955 to the present. These photographs cover agricultural and forest areas of the United States, and have a scale ranging from 1:15,840 to 1:80,000, primarily taken with black-and-white panchromatic film (Paine and Kiser, 2003). The National Agriculture Imagery Program (NAIP) has acquired imagery during the agricultural growing seasons in the continental United States since 2002. A primary goal of the NAIP program is to enable availability of digital orthophotography within a year of acquisition. By using the NAIP GIS Dataset Viewer (http://gis.apfo.usda.gov/gisviewer/), we can download the photographs by quarter quad in each state.

If a project requires using aerial photographs of a specific location during a defined period, then a flight mission should be considered by contracting with a commercial company. Some companies can do both flying and data processing. Many factors need to be considered in planning a flight mission (Paine and Kiser, 2003), including:

- Scale, flying height, and focal length
- Required percentage of sidelap and endlap
- Orientation, location, and the number of flight lines
- Total number of aerial photographs
- The amount of drift, crab, tilt, and image motion allowed
- Camera format, lens quality, shutter speed, aperture, inter-valometer setting, and cycling time
- Film and filter characteristics
- The season of year, time of day, and allowable cloud cover
- Hot spots
- Aircraft capacities (maximum and minimum speed, range, maximum altitude, load capacity, stability, and operating cost)

Paine and Kiser (2003) gave a detailed account for each of the above factors. One of the key factors is photo scale, which controls the amount of details we can see and interpret from the photos. The photo scale is normally determined after considering the objective of application, the size of the area to be covered, the topography of the region under investigation, and the funds available. Once the photo scale is set, focal length and the flying height of the aircraft must be matched [Eq. (4.1)]. From the relief displacement equation [Eq. (5.1)], it becomes clear that relief displacement would reduce as the flying height increases. Depending on the objective of application, we may want to minimize relief displacement or increase it. Photographs with little relief displacement are good for making maps, for example, orthophotographs; while those with larger amounts of relief displacement are more appropriate for viewing 3D effect and for measuring heights. The determination of flying height also needs to consider aircraft capacity, cost (flying higher costs more), and image motion.

Prior to any flight, the number of photographs and location of each photo must be planned to determine the number of flight lines. A flight line is a succession of overlapping aerial photographs taken along a single flight line. All flight lines are parallel to each other, so the airplane needs to make a 180° turn after finishing one flight line

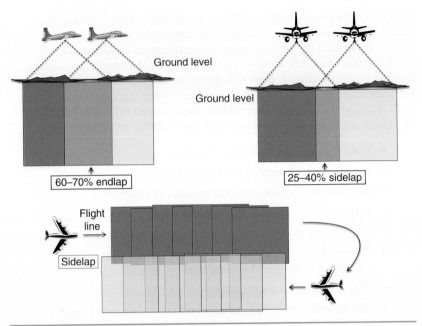

Figure 5.15   Endlap, sidelap, and flight line.

and return along the adjacent flight line (Fig. 5.15*a*). In developing a flight plan, normally 60 to 70% of endlap (also named forward over-lap) is designed. Endlap refers to the amount of overlap between two adjoining aerial photos in the same flight line, usually expressed as a percentage (Fig. 5.15*b*). The excessive overlapping of aerial photo-graphs in a flight line allows for photos viewed in three dimensions (3D) and facilitates taking measurements from the photos. Sidelap is the amount of overlap between two aerial photos in adjacent flight lines, ranging from 25 to 40% (Fig. 5.15*c*). Sidelap is designed so that we can obtain control points to make a mosaic of the photos for map-ping. Another consideration is to ensure no skipped area between flight lines. Sidelap is important because the actual location of a flight line and photo centers may be slightly altered due to errors caused by aircraft drifts, tilts, camera crabs (camera not being ori-ented parallel to the flight line), and other navigation problems (Paine and Kiser, 2003).

Both the season of the year and the time of day influence color/tonal contrasts of films and shadow length (Paine and Kiser, 2003). Sun angle changes with days and seasons, resulting in change in shadow length. In addition to sun angle, the decision for selecting a season for photographic flights must also consider features to be detected or mapped, the number of days suitable for aerial photography, and the film-filter combination to be used (Avery and Berlin, 1992). Because shadows cast by clouds can obscure many details on the ground, a photographic day is defined as one with 10% cloud cover or less (Avery and Berlin, 1992). To enable identification, interpretation, and mapping of the Earth's surface features, an important factor to be considered is the condition of vegetation cover. In the mid-latitude regions, deciduous trees and brush species turn color in the fall and lose their leaves in the winter. Projects interested in vegetation analysis therefore will need to schedule flights in the leaf-on season. Deciduous trees and brush species are frequently photographed in the summer, while remote sensing of conifer trees may take place in the early spring before deciduous brush leafs come out. Topographic, soil, lithological, and geological structure mapping are usually conducted in the leaf-off season, that is, in the spring or the fall when deciduous foliage and snow are absent. The winter is a good season for identifying big game animals.

The time of day is an important factor in aerial photography because the sun's illumination, controlled by solar angle, affects both the quantity of reflected light to the aerial camera and its spectral quality (Avery and Berlin, 1992). If shadow enhancement is not needed, photographic flights should be scheduled around noontime (specifically, within approximately 2 hours of local apparent noon); but if shadow enhancement is needed, then flights should be set for early morning or late afternoon, when shadows are longest because of low sun angle (Avery and Berlin, 1992). Early morning is preferred over late afternoon because early morning normally is less hazy and less turbulent (Avery and Berlin, 1992). Wrong timing (combination of season and the time of day) for photographic flights can result in hotspots, which appear as bleached out or overexposed portions of the photograph. A hotspot occurs when a straight line from the sun passes through the camera lens and intersects with the ground inside the photo ground coverage area and reflected light directly returns to the camera (Avery and Berlin, 1992). Hotspots tend to occur when photographing with high sun angle, low altitude, high flying height, and wide-angle camera lenses.

5.6 Stereoscopy

Humans with "normal" eyesight usually have stereoscopic vision, that is, the ability to see and appreciate the depth of field through the perception of parallax. Parallax is the apparent displacement of a viewed point or small object at a distance that results from a change in the point of observation (Short, 2009). When viewing an object, two eyes look from different vantages. The left eye sees a bit of the left side of the object, not seen by the right eye, while the right eye sees a bit of the right side, missed by the left eye. In this way, the

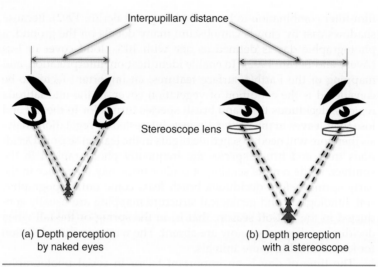

(a) Depth perception
     by naked eyes

(b) Depth perception
     with a stereoscope

FIGURE 5.16   Depth perception with naked eyes and a stereoscope.

human brain receives two images of that object, and fuses them into one image with the impression of depth (Fig. 5.16a). Each individual has a different degree of stereoscopic vision depending on a number of factors, for example, interpupillary distance, near or far sightedness, etc. The normal interpupillary distance in humans is 2.5 to 2.6 in. If this distance can be increased, we can then increase our perception of depth. Stereoscopes are designed to stretch greatly the interpupillary distance, so that we can perceive the exaggerated 3D photographic effect when viewing the stereo pairs of aerial photographs (Fig. 5.16b). The successive photo pairs were taken with approximately 60% overlap along a single flight line, with some of the scene in common within the pairs. When we position the stereo pair properly left/right and then view them through a stereoscope, the eye–brain reaction creates an impression of surface curvature or relief (i.e., depth perception), as though we are looking down from an airplane at the ground. We can experiment with viewing 3D effect without a stereoscope by looking at the pair of photos in Fig. 5.17 (Short, 2009). If you are successful, you should be able to see the strata (bands of rocks) appear to be inclined or "dipping" toward the upper right.

Figure 5.18a illustrates the proper procedure for using a pocket stereoscope to view a stereopair of aerial photographs in order to perceive the 3D effect. The photos must be properly aligned according to the flight line, and often one photo is placed on top of the other, which is then alternated to the top for viewing another area of overlap. The shadows should preferably be toward the viewer. The viewer should also keep the eye base and the long axis of the

FIGURE 5.17 A pair of stereo-images that many people may be able to view 3D effect with naked eyes.

stereoscope parallel to the flight line. It is usually possible to see the 3D effect in the photos, but it may take a while to practice. A major difference between normal viewing and stereo viewing lies in depth perception. Whereas normal viewing of an aerial photograph on a table requires us to focus on the surface level of the photograph, stereo viewing asks us to focus our eyes at infinity, meaning that the line of sight for one eye is parallel to the other. Figure 5.18*b* illustrates another type of device for stereovision. The mirror stereoscope is more expensive than a pocket stereoscope, but it is often easier to use and allows a larger area to be seen in 3D. The distance between two photos may be adjusted by moving the left photo to the left and the right photo to the right until seeing the 3D effect. No matter what type of device is used, pseudo stereo should be avoided, which can be caused by erroneous reversal of prints or by viewing photos with shadows falling away from rather than toward the interpreter. In areas of high vertical relief, scale differences in

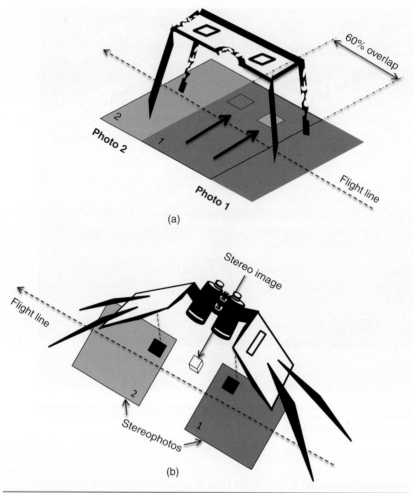

FIGURE 5.18   Use of a pocket steoreoscope (*a*) and a mirror stereoscope (*b*) to view 3D photographic effect (*After Arnold, 2004*).

adjacent photographs may make it difficult to obtain a 3D effect (Avery and Berlin, 1992).

5.7   Orthophotography

An orthophotograph is the reproduction of an aerial photograph with all tilts and relief displacements removed and with a uniform scale over the whole photograph. As discussed in Section 5.4, a regular vertical photograph uses central projection while maps use an orthogonal projection. The production of orthophotographs requires the transformation of central projection into orthogonal projection to make aerial photographs "planimetric." The resulting planimetric maps have all features on the ground surface portrayed at their correct horizontal positions, so that reliable measurement can be made from these "photomaps." An orthoimage is the digital version of an orthophotograph, which can be produced from a stereoscopic pair of scanned aerial photographs or from a stereopair of satellite images (Lo and Yeung, 2002). The production of an orthophotograph or orthoimage requires the use of a DEM to register properly to the stereo model to provide the correct height data for differential rectification of the image (Jensen, 2005). Orthoimages are increasingly used to provide the base maps for GIS databases upon which thematic data layers are overlaid (Lo and Yeung, 2002).

The USGS produces three types of digital orthophoto quadrangle (DOQ) products. The DOQ combines the image characteristics of the original photograph with the same positional accuracy and scale of a topographic map. DOQs are in black-and-white, natural color, or color-IR images with 1-m ground resolution. Figure 5.19 show a 3.75-minute black-and-white DOQ and a 3.75-minute color-IR DOQ for Washington, D.C.

The types of DOQs by the USGS include:

1. 3.75-minute (quarter-quad) DOQs cover an area measuring 3.75-minute longitude by 3.75-minute latitude. Most of the United States is currently available.

Figure 5.19 A 3.75-minute black-and-white digital orthophoto quadrangle (DOQ) and a 3.75-minute color-IR DOQ for Washington, D.C. (*Courtesy of U.S. Geological Survey.*)

2. 7.5-minute (full-quad) DOQs cover an area measuring 7.5-minute longitude by 7.5-minute latitude. Full-quad DOQs are mostly available for Oregon, Washington, and Alaska. Limited coverage may also be available for other states.

3. Seamless DOQs are available for free download from the USGS Website (http://seamless.usgs.gov/). DOQs on this site are the most current version and are available for the conterminous United States.

5.8   Summary

Although aerial photographs can be taken vertically or obliquely, vertical photos are commonly used for taking measurements and making maps. An aerial photograph uses a central perspective projection, causing an object in the Earth's surface to be displaced away from the optical center (often overlapping with the geometric center) of the photograph depending on its height and location in the photograph. This relief displacement makes it possible to determine mathematically the height of the object by using a single photograph. Alternatively, we can use a pair of photographs to measure the height of an object. Such a pair of photographs must be taken with a 50 to 60% endlap to ensure stereoscopic coverage and stereo viewing. When obtaining vertical aerial photographs, the aircraft normally flies in a series of flight lines with an overlap (i.e., sidelap) of 25 to 40%, so that we will not skip any area under investigation and can obtain control points to create a mosaic of the photos for mapping. Planning a flight mission is a complex process, involving a number of factors such as the project objective, camera and film, time and season, aircraft capacities and movement, and image motion.

Recent advances in computer and imaging technologies have transformed the traditional analog photogrammetry into digital (softcopy) photogrammetry, which uses modern technologies to produce accurate topographic maps, orthophotographs, and orthoimages by using the principles of photogrammetry. Orthoimages are increasingly used to provide the base maps for GIS databases upon which thematic data layers are overlaid (Lo and Yeung, 2002).

Key Concepts and Terms

Absolute parallax   The sum of the differences between the base of conjugate objects and their respective nadirs. This is always measured parallel to the air base.

Air base   The line segment joining the principal point and the conjugate principal point constitute the flight line of the aircraft, also called base line.

Conjugate principal points   The principal point can be transferred stereoscopically onto the adjacent (left and right) photographs of the same flight line by pricking through its transferred positions. These transferred points are called conjugate principal points.

Differential parallax   The difference in the absolute stereoscopic parallax at the top and the base of the object on a pair of photographs.

Endlap   The amount of overlap between two aerial photos in the same flight line, usually expressed as a percentage.

Fiducial marks   Marks that appear on each aerial photo in the sides, corners, or both, and are used to determine the precise location of the principal point. If the x-axis is arbitrarily assigned to the imaginary flight line direction on the photograph and the y-axis to a line perpendicular to the x-axis, the two axes correspond to the lines connecting the opposite fiducial marks.

Isocenter   The point halfway between the principal point and the nadir and on the line segment joining these two points on the photograph. It is a point intersected by the bisector of the angle between the plumb line and the optical axis.

Nadir   Also called vertical point or plumb point. The nadir is the intersecting point between the plumb line directly beneath the camera center at the time of exposure and the ground.

Oblique aerial photographs   Photos are taken with the camera pointed to the side of the aircraft. The ground area imaged by oblique photography is trapezoidal in shape. Oblique aerial photographs can be divided further into two types—low oblique and high oblique. High-oblique photographs usually include the ground surface, horizon, and a portion of sky, while low-oblique photographs do not show horizon.

Orthoimage   The digital version of an orthophotograph. It can be produced from a stereoscopic pair of scanned aerial photographs or from a stereopair of satellite images.

Orthophotograph   An orthophotograph is the reproduction of an aerial photograph with all tilts and relief displacements removed and a constant scale over the whole photograph.

Parallax   Also referring to image displacement. This is the apparent displacement of the position of an object, caused by a shift in the point of observation. Parallax occurs in any vertical aerial photo of ground features that lie above or below the average ground elevation.

Principal point   The optical or geometric center of a photograph.

Sidelap   The amount of overlap between two aerial photos in adjacent flight lines, usually expressed as a percentage.

Stereogram   A stereopair or stereotriplet of aerial photos mounted for proper stereovision.

Stereopair   The successive photo pairs taken with approximately 60% overlap along a single flight line, with some of the scene in common within the pairs.

Vertical aerial photographs   Photographs are acquired with the camera pointed vertically down (or slightly tilted, less than 3° from the vertical) to the Earth's surface. The ground area imaged by vertical photography is square.

Review Questions

1.   What are the benefits for using vertical aerial photographs? Suggest circumstances when oblique aerial photographs are more appropriate than vertical aerial photographs.

2.   If an aerial photo were taken from an altitude of 3000 feet with an aerial camera that has a 6-in. focal length, what would the scale of the photo be? If a building on this photo is measured to have a photo displacement from top to bottom of 0.5 in., and the radial distance from the photo nadir to the top of the building is 3 in., what would the height of the building be?

3.   Explain why the time of day and the season of year are important in planning a photographic flight mission.

4.   What are the differences between an aerial photograph and an orthophotograph? How is an orthophotograph generated?

5.   How does a stereoscope help us to see aerial photographs in 3D?

6.   How does a pseudoscopic view happen?

References

Arnold, R. H. 2004. *Interpretation of Airphotos and Remotely Sensed Imagery*. Long Grove, Ill.: Waveland Press.

Avery, T. E., and G. L. Berlin. 1992. *Fundamentals of Remote Sensing and Airphoto Interpretation*, 5th ed. Upper Saddle River, N.J.: Prentice Hall.

Campell, J. B. 2007. *Introduction to Remote Sensing*, 4th ed. New York: The Guilford Press.

Colwell, R. N. 1997. History and place of photographic interpretation. In Philipson, W. R. *Manual of Photographic Interpretation*, 2nd ed. Bethesda, Md.: American Association of Photogrammetry and Remote Sensing.

Fahsi, A. 1996. Lecture 6: Geometry of Aerial Photography. In *Remote Sensing Core Curriculum Volume 1, Air Photo Interpretation and Photogrammetry*. http://userpages.umbc.edu/~tbenja1/umbc7/santabar/rscc.html (accessed September 10, 2010).

Jensen, J. R. 2005. *Introductory Digital Image Processing: A Remote Sensing Perspective,* 3rd ed. Upper Saddle River, N.J.: Pearson Prentice Hall.

Lo, C. P., and A. K. W. Yeung. 2002. *Concepts and Techniques of Geographic Information Systems.* Upper Saddle River, N.J.: Prentice Hall.

Paine, D. P., and J. D. Kiser. 2003. *Aerial Photography and Image Interpretation.* Hoboken, N.J.: John Wiley & Sons.

Short, N. M., Sr. 2009. *The Remote Sensing Tutorial,* http://rst.gsfc.nasa.gov/(accessed February 26, 2010).

Wolf, P. R., and B. A. Dewitt. 2000. *Elements of Photogrammetry with Applications in GIS.* New York: McGraw-Hill.

CHAPTER 6

Remote Sensors

6.1 Introduction

In Chap. 3, we briefly described the properties of satellite imagery. In this chapter, we extend our discussion to all non-photograph images, and focus on how these images are acquired physically with various remote sensors. Remote sensing systems or sensors can be grouped into two types, that is, passive and active sensors. As discussed previously, the main source of energy for remote sensing comes from the sun. Remote sensors record solar radiation reflected or emitted from the Earth's surface. When the source of the energy comes from outside a sensor, it is called a passive sensor. Examples of passive sensors include photographic cameras, optical-electrical sensors, thermal infrared (IR) sensors, and antenna sensors. Because passive sensors use naturally occurring energy, they can only capture data during the daylight hours. The exception is thermal IR sensors, which can detect naturally emitted energy day or night, as long as the amount of energy is large enough to be recorded. We begin our discussion with a concise comparison between optical-electrical sensors and photographic cameras, which is followed by a more detailed treatment of the basic processes and operating principle of across-track and along-track scanners, two commonly used optical-electrical sensors for acquiring multispectral imagery. We then discuss thermal IR sensors before turning to active sensors.

Active sensors use the energy source coming from within the sensor. They provide their own energy that is directed toward the target to be investigated. The energy scattered back or reflected from that target is then detected and recorded by the sensors. An example of active sensing is radar (radio detection and ranging), which transmits a microwave signal toward the target and detects and measures the backscattered portion of the signal. Another example is Lidar (light detection and ranging). It emits a laser pulse and precisely measures its return time to calculate the height of each target. Active sensors can be used to image the surface at any time, day or night, and in any season. Active sensors can also be used to examine wavelengths that are not sufficiently provided by the sun, such as microwaves, or to better control the way a target is illuminated (Canada Centre for Remote Sensing, 2007).

6.2   Electro-Optical Sensors

We have learned that satellite images are acquired by electro-optical sensors. Each sensor may consist of various amounts of detectors. The detectors receive and measure reflected or emitted radiance from the ground as electrical signals, and convert them into numerical values, which are stored onboard or transmitted to a receiving station on the ground. The incoming radiance is often directed toward different detectors, so that the radiance within a specific range of wavelength can be recorded as a spectral band. The major types of scanning systems employed to acquire multispectral image data include across-track and along-track scanners.

Electro-optical sensors offer several advantages over photographic cameras. First, while the spectral range of photographic systems is restricted to the visible and near-IR regions, electro-optical sensors can extend this range into the whole reflected IR and the thermal IR regions. Electro-optical sensors are capable of achieving a much higher spectral resolution. Because electro-optical sensors acquire all spectral bands simultaneously through the same optical system, it is also much easier to register multiple images (Canada Centre for Remote Sensing, 2007). Second, some electro-optical sensors can detect thermal IR energy in both daytime and nighttime mode (Avery and Berlin, 1992). Third, since electro-optical sensors record radiance electronically, it is easier to determine the specific amount of energy measured, and to convert the incoming radiance over a greater range of values into digital format (Canada Centre for Remote Sensing, 2007). Therefore, electro-optical sensors can better utilize radiometric resolution than photographic cameras. Fourth, with photographic cameras we have to use films as both the detector and storage medium, whereas electro-optical detectors can be used continuously, making the detection process renewable (Avery and Berlin, 1992). Fifth, some electro-optical sensors employ an in-flight display device, allowing a ground scene to be viewed in a near real-time mode (Avery and Berlin, 1992). Finally, the digital recording in electro-optical sensors facilitates transmission of data to receiving stations on the ground and immediate processing of data by computers (Canada Centre for Remote Sensing, 2007). Of course, we should note that aerial photographs by cameras have the clear advantage of recording extremely fine spatial details.

6.3   Across-Track Scanning Systems

Instead of imaging a ground scene entirely in a given instant in time (as with a photographic camera), across-track scanners scan (i.e., image) the Earth's surface in a contiguous series of narrow strips by using a rotating/oscillating mirror (Fig. 6.1). These ground strips are oriented perpendicular to the flight path of the platform (e.g., satellite). As the plat moves forward, new ground strips are created. Successive ground strips build up a 2D image of the Earth's surface along the flight line. This technique is sometimes called "whiskbroom" scanning because the scanner "whisks" the Earth's surface as the platform advances along the flight direction. The incoming reflected or emitted radiation is frequently separated into several spectral bands, as these remote sensing scanning systems are capable of multispectral imaging. The near UV, visible, reflected IR, and thermal IR radiation can all be used in a multispectral remote sensing system, which usually has 5 to 10 spectral bands. A set of electronic detectors, which are sensitive to a specific range

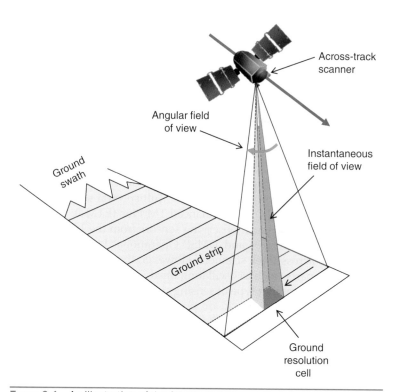

FIGURE 6.1   An illustration of the functioning of an across-track or whiskbroom scanner.

of wavelengths, detects and measures the incoming radiation for each spectral band and saves it as digital data.

For a given altitude of the platform, the angular field of view, that is, the sweep of the mirror, determines the width of the imaged swath. Airborne scanners typically sweep large angles (between 90° and 120°), while satellites need only to sweep fairly small angles (10° to 20°) to cover a broad region due to their higher altitudes. The instantaneous field of view (IFOV) of the sensor determines the ground area viewed by the sensor in a given instant in time. This ground area, known as a ground resolution cell, is subjected to the influence of the altitude of the platform. As discussed in Chap. 3, IFOV defines the spatial resolution of a sensor. Because the distance from the sensor to the target increases toward the edges of the swath, the ground resolution cells are larger toward the edges than at the center of the ground swath (i.e., nadir). This change in the sensor-target distance gives rise to a scale distortion in the acquired images, which is often corrected during the stage of image processing before providing to the users. The length of time that the IFOV "sees" a ground resolution cell as the rotating mirror scans is called the dwell time. It is generally quite short, but it influences the design of the spatial, spectral, and radiometric resolution of the sensor (Canada Centre for Remote Sensing, 2007). A small IFOV, in spite of being associated with high spatial resolution, limits the amount of radiation received by the sensor, whereas a larger IFOV would receive more radiation, allowing detection of small variations in reflectance or emittance. Moreover, a larger IFOV generates in effect a longer dwell time over any given area, and improves the radiometric resolution. A large IFOV also has a high signal-to-noise ratio (SNR; signal greater than background noise).

Across-track scanners generally have a wide image swath. Although their mechanical system is complex, their optical system is relatively simple. The across-track scanners have been widely used in satellite remote sensing programs. Well-known examples include those scanners onboard the Landsats: Multispectral scanner (MSS), Thematic Mapper (TM), and Enhanced TM Plus (ETM+). These sensors were the prime Earth-observing sensors from the 1970s into the 1980s. NOAA's AVHRR and DMSP Operational Linescan Systems also use the across-track scanners. However, these instruments contain moving parts, such as oscillating mirrors, which are easily worn out and failed. Another type of sensing systems was developed in the interim, namely along-track, or pushbroom, scanners, which use charge-coupled devices (CCDs) as the detector.

6.4   Linear-Array (Along-Track) Scanning Systems

Like across-track scanners, along-track scanners also use the forward motion of the platform to record successive scan lines and build up a 2D image. The fundamental difference between the two types of

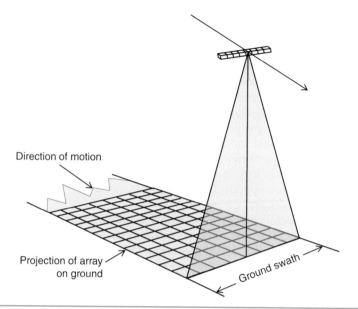

Direction of motion

Projection of array
on ground

Ground swath

FIGURE 6.2   An illustration of the functioning of an along-track or pushbroom scanner.

scanning systems lies in the method in which each scan line is built. Along-track scanners detect the field of view at once by using a linear array of detectors, which are located at the focal plane of the image formed by the lens systems (Fig. 6.2). Linear arrays may consist of numerous CCDs positioned end to end. The sensor records the energy of each scan line simultaneously as the platform advances along the flight track direction, which is analogous to "pushing" the bristles of a broom along a floor. For this reason, along-track scanning systems are also known as pushbroom scanners. Each individual detector measures the energy for a single ground resolution cell, and thus the IFOV of a single detector determines the spatial resolution of the system. Each spectral band requires a separate linear array. Area array can also be used for multi-spectral remote sensing, with which a dispersing element is often used to split light into narrow spectral bands.

Along-track scanners offer several advantages over across-track scanners. Because pushbroom scanners have no moving parts (oscillating mirrors), their mechanical reliability can be relatively high. Moreover, along-track scanners allow each detector to "see" and measure the energy from each ground resolution cell for a longer period of time (dwell time), and therefore enables a stronger signal to be recorded and a higher signal-to-noise ratio, as well as a greater range in the signal, which improves the radiometric resolution of the system. Further, the increased dwell time makes it possible for smaller IFOVs and narrower bandwidths for each detector. Therefore, along-track scanners can have finer spatial and spectral resolution without sacrificing the radiometric resolution. In addition, the geometric accuracy in the across track direction is higher owing to the fixed relationship among the detector's elements. Because detectors are usually solid-state microelectronic devices, they are generally smaller and lighter, and require less power. These advantages make it well suited for along-track scanners to be onboard small satellites and airplanes (Avery and Berlin, 1992). A disadvantage of linear array systems is the need for cross-calibrating thousands of detectors to achieve uniform sensitivity across the array. Another limitation of the pushbroom technology is that current commercially available CCD detectors cannot operate in the mid-IR and longer wavelengths (Lillesand, Kiefer, and Chipman, 2008).

CCD detectors are now commonly employed on air- and spaceborne sensors. The first airborne pushbroom scanner is the multispectral electro-optical imaging scanner (MEIS) built by the Canada Centre for Remote Sensing. It images in eight bands from 0.39 to 1.1 μm (using optical filters to produce the narrow band intervals) and uses a mirror to collect fore and aft views (along-track) suitable as stereo imagery. The German Aerospace Research Establishment (DFVLR) developed the first pushbroom scanner to be flown in space. The modular optico-electronic multispectral scanner (MOMS) was aboard shuttle mission STS-7 and STS-11 in 1983 and 1984. It images two bands at 20 m resolution at the wavelength range of 0.575 to 0.625 μm and 0.825 to 0.975 μm. The first use of a CCD-based pushbroom scanner on an Earth-observing satellite was on the French SPOT-1 launched in 1986 with the sensor SPOT-HRV. Many other satellite remote sensing programs have now adopted the technology of along-track scanners, including the Indian satellite IRS, Japanese ALOS, Chinese HJ-1 series satellites, U.S. EO-1 (ALI sensor), QuickBird, and IKONOS.

6.5   Thermal IR Sensors

Thermal IR scanners are a special type of electro-optical sensor. The thermal scanners detect and measure the thermal IR portion of electromagnetic radiation, instead of visible, near-IR and reflected IR radiations. As discussed in Chap. 2, atmospheric absorption is very strong in some portions of the electromagnetic spectrum due to the existence of water vapor and carbon dioxide. As a result, thermal IR sensing is limited to two specific regions from 3 to 5 μm and from 8 to 14 μm, known as atmospheric windows. Comparing with the visible and reflected IR spectra, thermal IR radiation is much weaker, which requires the thermal sensors to have large IFOVs sufficient to detect the weak energy to make a reliable measurement. The spatial resolution of thermal IR channels is usually coarser than that for the visible and reflected IR bands. For example, Landsat TM sensor has a spatial resolution of 120 m in its thermal IR band, while from visible to mid-IR bands the resolution is 30 m. An advantage of thermal imaging is that images can be acquired at both day and night time, as long as the thermal radiation is continuously emitted from the terrain. Photo detectors sensitive to the direct contact of photons on their surface are used in the thermal sensors to detect emitted thermal IR radiation, and are made from combinations of metallic materials. To sense thermal IR radiation in the 8- to 14-µm interval, the detector is usually an alloy of mercury-cadmium-tellurium (HgCdTe). Also used for this thermal imaging window is mercury-doped germanium (Ge:Hg), which is effective over a broader spectral range of 3 to 14 µm. Indium-antimony (InSb) is the alloy used in detectors in the range of 3 to 5 µm. To maintain maximum sensitivity, these detectors must be cooled onboard to keep their temperatures between 30 and 77 K, depending on the detector type. The cooling agents, such as liquid nitrogen or helium, are contained in a vacuum bottle named "Dewar" that encloses the detector. Some spacecraft designs with radiant cooling systems take advantage of the cold vacuum of outer space (Short, 2009). The high sensitivity of the detectors ensures that fine differences in the emitted radiation can be detected and measured, leading to high radiometric resolution. A better understanding of this radiometric concept is via SNR. This signal is an electrical current related to changes in the incoming radiant energy from the terrain, while noise refers to unwanted contribution from the sensor (Campbell, 2007). Higher sensitivity means to improve the SNR by limiting the thermal emissions of the sensor while keeping a stable, minimum magnitude of the signal from the feature imaged.

In addition to thermal detectors, thermal scanners have two other basic components: an optical-mechanical scanning system and an image recording system (Fig. 6.3). For across-track scanners, an electrical motor is mounted underneath the aircraft, with the rotating shaft oriented parallel to the aircraft fuselage—thus, the flight direction. The total angular field of view typically ranges from 90° to 120°, depending on the type of sensor (Jensen, 2007). The incoming thermal IR energy is focused from the scan mirror onto the detector, which converts the radiant energy into an appropriate level of analog electrical signal. The resulting electrical signals are sequentially amplified (due to the low level of thermal IR energy from the terrain) and recorded on magnetic tapes or other devices. These signals can be displayed in monitors, and can be recorded and analyzed digitally after analog-to-digital conversion. For each scan, the detector response must be calibrated to obtain a quantitative expression of radiant temperatures. The calibrated thermal scanners usually have

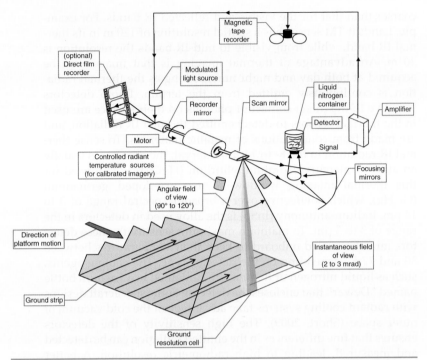

FIGURE 6.3   An illustration of the functioning of a thermal IR scanner (*After Sabins, 1997*).

two internal temperature references (electrical elements, mounted at the sides of the scanning mirror), one as the cold reference and the other as the warm reference. The two references provide a scale for determining the radiant temperatures of ground objects (Avery and Berlin, 1992).

Recent technological advances in CCDs make it possible for thermal IR imaging based on area and linear arrays of detectors (Jensen, 2007). As with other along-track scanners, thermal IR area/linear arrays permit a longer dwell time, and thus improve radiometric resolution and the accuracy of temperature measurements. Moreover, because the alignment of all detectors is fixed, the geometric fidelity of thermal IR imagery is much enhanced. Some forward-looking IR (FLIR) systems, used in military reconnaissance and law enforcement, adopt area-array technology. The Canadian firm, Itres (http://www.itres.com/), has built a thermal IR sensor based on linear-array technology. The newest type of the sensor, TABI-1800 (thermal airborne broadband imager), measures thermal IR spectra in the range of 8 to 12 μm with a swath of 320 to 1800 pixels. It can provide a temperature resolution up to 1/10 of a degree.

6.6   Passive Microwave and Imaging Radar Sensors

The electromagnetic radiation in the microwave wavelength region (1 cm to 1 m in wavelength) has been widely used in remote sensing of the Earth's land surface, atmosphere, and ocean. In comparison with visible, reflected, and thermal IR spectra, microwave has a longer wavelength. This property permits microwave to penetrate cloud cover, haze, dust, etc. in the atmosphere, while not to be affected by atmospheric scattering, which affects shorter optical wavelengths. The detection of microwave energy under almost all weather and environmental conditions is especially important for tropical areas, where large cloud covers occur frequently throughout the year.

Passive microwave sensors operate in a similar manner as those for thermal IR imaging. The sensors are typically radiometers, which require large collection antennas, either fixed or movable. The passive microwave sensors detect and measure naturally emitted microwave energy within their field of view, with wavelengths between 0.15 and 30 cm. The emitted energy is controlled mainly by the temperature and moisture properties of the emitting object or surface. Because the wavelengths are so long, this emitted energy is small compared to optical wavelengths (visible and reflected IR). As a result, the fields of view for most passive microwave sensors are large in order to detect enough energy to record a signal, and the data acquired by such sensors are characterized by low spatial resolution.

Passive microwave remote sensing using air- and space-borne sensors has found uses for several decades in the studies of terrestrial systems, meteorology, hydrology, and oceanography. Owing to the sensitivity of microwaves to moisture content, passive microwave sensing is widely used for detecting soil moisture and temperature, as well as assessing snowmelt conditions. In addition, microwave radiation from below thin soil cover provides helpful information about near-surface bedrock geology. Onboard some meteorological satellites, passive microwave sensors are the devices used to measure atmospheric profiles, water and ozone content in the atmosphere, and to determine precipitation conditions. On oceans, passive microwave sensors track sea ice distribution and currents, and assess sea surface temperature and surface winds. These sensors are also useful in monitoring pollutants in the oceans and in tracking oil slicks.

Active microwave sensors provide their own source of microwave radiation to illuminate the target. Active microwave sensors can be imaging or non-imaging. The most common form of imaging active microwave sensor is radar. Radar typically consists of four components: a transmitter, a receiver, an antenna, and an electronics system to process and record the data. Radar measures both the

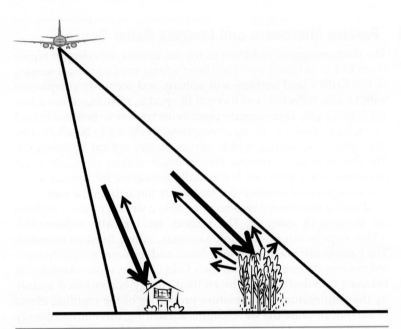

FIGURE 6.4   Imaging radar pulse transmittance and backscattering from the Earth's surface targets.

strength and roundtrip time of microwave signals that are emitted by a radar antenna and reflected off a target (object or surface; Fig. 6.4). The radar antenna alternately transmits and receives pulses at particular microwave wavelengths and polarizations, the orientation of the electric field (Freeman, 1996). When reaching the Earth's surface, the energy in the radar pulse may be scattered in all directions, possibly with some reflected back toward the antenna. This backscattered back energy is then reached by the radar and received by the antenna in a specific polarization. These radar echoes are converted to digital data and passed to a data recorder for later processing and display as an image (Freeman, 1996). Because the radar pulse travels at the speed of light, it is possible to use the measured time for the roundtrip of a particular pulse to calculate the distance or range to the reflecting object (Freeman, 1996).

Imaging radar sensing is apparently different from optical and thermal remote sensing. As an active sensor, radar can be used to image any target at any time, day or night. This property of radar, when combined with microwave energy's ability to penetrate clouds and most rain, makes it an anytime and all-weather sensor. Figure 6.5 illustrates a spaceborne imaging radar-C/X-band synthetic aperture radar (SIR-C/X-SAR) color composite image of southeast Tibet on April 10, 1994. It was created by assigning three SIR-C/X-SAR bands,

Lhasa River

FIGURE 6.5   Spaceborne imaging radar-C/X-band synthetic aperture radar (SIR-C/X-SAR) of southeast Tibet on April 10, 1994, onboard the space shuttle Endeavour. The image covers a rugged mountainous area of southeast Tibet, approximately 90 km (56 mi) east of the city of Lhasa. Mountains in this area reach approximately 5800 m (19,000 feet) above sea level, while the valley floors lie approximately 4300 m (14,000 feet). SIR-C/X-SAR, a joint mission of the German, Italian, and U.S. space agencies, is part of NASA's Mission to Planet Earth. (*Image courtesy of NASA/JPL.*)

L-band (24 cm), C-band (6 cm), and X-band (3 cm), as RGB. The SIR-C/X-SAR is part of NASA's Mission to Planet Earth. The multi-frequency data are used by the international scientific community to better understand the global environment and how it is changing. Geologists use radar images like the one in Fig. 6.5 to map the distribution of different rock types and to understand the history of the formation and erosion of the Tibetan Plateau. Figure 6.6 shows a SIR-C/X-SAR image of Athens, Greece on October 2, 1994, illustrating its urban landscapes and surrounding areas. Densely populated urban areas appear in pink and light green. Numerous ships, shown as bright dots, can be seen in the harbor areas in the upper left part of the image. The unique characteristics and applications of imaging radar will be discussed in detail in Chap. 9.

Non-imaging microwave sensors include altimeters and scatterometers. Radar altimeters send out microwave pulses and measure the roundtrip time delay to the targets. The height of the surface can be determined based on the time delay of the return signals. Altimeters look straight down at the nadir below the platform (aircraft or satellite) to measure heights or elevations (Canada Centre for Remote Sensing, 2007). Radar altimetry is commonly used on aircrafts for height determination, and for topographic

Acropolis

Piraeus

FIGURE 6.6   SIR-C/X-SAR image of Athens, Greece on October 2, 1994. The colors are assigned to different radar frequencies and polarizations as follows: red to L-band (horizontally transmitted and received), green to L-band (horizontally transmitted and vertically received), and blue to C-band (horizontally transmitted and received). (*Image courtesy of NASA/JPL.*)

mapping and sea surface height estimation with the platforms of aircrafts and satellites. Another type of non-imaging sensor is a scatterometer. Scatterometers are used to make reliable measurements of the amount of backscattered energy, which is largely influenced by the surface properties and the angle at which the microwave energy strikes the target (Canada Centre for Remote Sensing, 2007). For example, a wind scatterometer is used to measure wind speed and direction over the ocean surface (Liew, 2001). It sends out pulses of microwaves in multiple directions and measures the magnitude of the backscattered signal, which is related to the ocean surface roughness. The information about the surface roughness can be used to derive the wind speed and direction of the ocean's surface (Liew, 2001).

6.7   Lidar

Lidar is a technology that dates back to the 1970s, when an airborne laser scanning system was developed. Like radar, it is an active remote sensing technology. The first system was developed by NASA, and it operated by emitting a laser pulse and precisely measuring its return time to calculate the range (height) by using the speed of light.

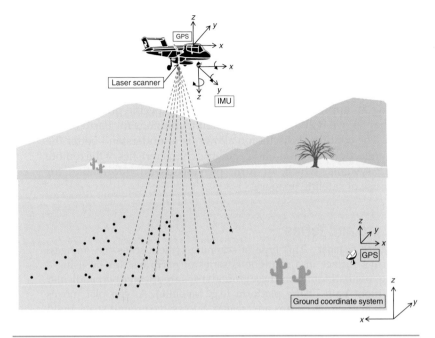

FIGURE 6.7   An illustration of the functioning of a Lidar system.

Later, with advent of the global positioning system (GPS) in the late 1980s and rapidly pulsing laser scanners, the necessary positioning accuracy and orientation parameters were achieved. The Lidar system has now become comprehensive, typically with four primary components, including a rapidly pulsing laser scanner, precise kinematic GPS positioning, inertial measurement units (IMUs) for capturing the orientation parameters (i.e., tip, tilt, and roll angles), and a precise timing clock (Renslow, Greenfield, and Guay, 2000). With all the components, along with software and computer support, assembled in a plane, and by flying along a well-defined flight plan, Lidar data are collected in the form of records of the returned-pulse range values and other parameters (e.g., scanner position, orientation parameters, etc.). After filtering noise and correcting, laser point coordinates are calculated, and a basic ASCII file of x, y, and z values for Lidar data is formed, which may be transformed into a local coordinate system. Figure 6.7 illustrates how Lidar works.

6.8   Summary

Based on the way reflected or emitted energy is used and measured during sensing, remote sensors are most commonly grouped into two categories: passive and active sensors. Passive sensors are subgrouped further according to the data recording method, and include across-track, along-track, and thermal IR scanners. Active sensors are further broken down into radar and Lidar sensors, which have distinct data collection methods and the geometric properties in resultant images. In this chapter, we examined the mechanism and process that each type of sensor acquires digital imagery. This knowledge is important for image interpretation and analysis, as well as for selecting an appropriate type of sensor for a specific project. In the next chapter, we will discuss main satellite missions for Earth observation. The continued discussion allows us to understand better the image collection processes and geometric characteristics of the remote sensing systems, which, in turn, provide useful information for proper use of satellite imagery.

Key Concepts and Terms

Across-track scanner   This type of scanning system scans (i.e., images) the Earth's surface in a contiguous series of narrow strips by using a rotating/oscillating mirror. These ground strips are oriented perpendicular to the flight path of the platform (e.g., satellite). As the plat moves forward, new ground strips are created. Successive ground strips build up a 2D image of the Earth's surface along the flight line. This technique is sometimes called "whiskbroom" scanning. The incident radiation is frequently separated into several spectral bands, as these scanning systems are capable of multispectral imaging.

Active sensor   Provides its own energy that is directed toward the target to be investigated. The energy scattered back or reflected from that target is then detected and recorded by the sensors, such as radar and Lidar.

Along-track scanner   Detects the field of view at once by using a linear array of detectors, which are located at the focal plane of the image formed by the lens systems. Linear arrays may consist of numerous charge-coupled devices positioned end-to-end. The sensor records the energy of each scan line simultaneously as the platform advances along the flight track direction. Along-track scanning systems are also known as pushbroom scanners. Each spectral band requires a separate linear array. Area array can also be used for multi-spectral remote sensing.

Angular field of view   The sweep of the mirror used in a scanning system. For a given altitude of the platform, it determines the width of the imaged swath.

Dwell time   The length of time that the IFOV observes a ground resolution cell as the rotating mirror scans.

Electro-optical sensor   Detector that receives and measures reflected or emitted radiance from the ground, as electrical signals, and converts them into numerical values for display as 2D images. Examples of electro-optical sensors include vidicon camera, video camera, across-track scanner, and along-track scanner.

IMUs   Inertial measurement units for capturing the orientation parameters (i.e., tip, tilt, and roll angles).

Instantaneous field of view (IFOV)   A measure of the ground area viewed by a single detector element in a given instant in time.

Lidar   An acronym for light detection and ranging. Lidar is an active remote sensing technology, and is operated by emitting a laser pulse and precisely measuring its return time to calculate the range (height) by using the speed of light.

Passive sensor   When the source of the energy comes from outside a sensor, it is called a passive sensor. Remote sensors that record solar radiation reflected or emitted from the Earth's surface belong to passive sensors, such as photographic cameras, optical-electrical sensors, thermal IR sensors, and antenna sensors.

Radar   An acronym of radio detection and ranging. Radar typically consists of four components: a transmitter, a receiver, an antenna, and an electronics system to process and record the data.

Radar altimeter   Sends out microwave pulses and measures the roundtrip time delay to the targets. The height of the surface can be determined based on the time delay of the return signals. Radar altimetry is commonly used on aircrafts for height determination, and for topographic mapping and sea surface height estimation with the platforms of aircrafts and satellites.

Scatterometer   Used to make reliable measurements of the amount of backscattered energy, which is largely influenced by the surface properties and the angle at which the microwave energy strikes the target. For example, a wind scatterometer is used to measure wind speed and direction over the ocean's surface.

Signal-to-noise (S/N) ratio   An indicator of radiometric quality. In thermal remote sensing, the signal is an electrical current related to changes in the incoming radiant energy from the terrain, while noise refers to unwanted contribution from the sensor.

Thermal infrared scanner   A special type of electro-optical sensor that detects and measures only the thermal IR portion of electromagnetic radiation.

Review Questions

1. What advantages do electro-optical sensors possess over photographic cameras? Are there any disadvantages?

2. Explain the principle of across-track scanning and how it differs from along-track scanning. How do the differences in the scanning method affect the geometric and radiometric quality of imagery?

3. Compared to other types of electro-optical sensors, what is special about a thermal IR scanner? How does it work?

4. Passive microwave sensing and radar have some commonalities as well as critical differences. Can you justify this statement with your reasons?

5. Identify and illustrate some application areas for non-imaging microwave sensors, such as altimeters and scatterometers.

6. Can you explain the principle of Lidar?

References

Avery, T. E., and G. L. Berlin. 1992. *Fundamentals of Remote Sensing and Airphoto Interpretation*, 5th ed. Upper Saddle River, N.J.: Prentice Hall.

Campbell, J. B. 2007. *Introduction to Remote Sensing*, 4th ed. New York: Guilford Press.

Canada Centre for Remote Sensing. 2007. *Tutorial: Fundamentals of Remote Sensing*, http://www.ccrs.nrcan.gc.ca/resource/tutor/fundam/chapter2/08_e.php. Accessed on December 3, 2010.

Freeman, T. 1996. *What* is Imaging Radar? http://southport.jpl.nasa.gov/. Accessed on January 23, 2011.

Jensen, J. R. 2007. *Remote Sensing of the Environment: An Earth Resource Perspective*, 2nd ed. Upper Saddle River, N.J.: Prentice Hall.

Liew, S. C. 2001. *Principles of Remote Sensing.* A tutorial as part of the "Space View of Asia, 2nd Edition" CD-ROM, the Centre for Remote Imaging, Sensing and Processing (CRISP), the National University of Singapore. http://www.crisp. nus.edu.sg/~research/tutorial/rsmain.htm. Accessed January 25, 2011.

Lillesand, T. M., R. W. Kiefer, and J. W. Chipman. 2008. *Remote Sensing and Image Interpretation*, 6th ed. Hoboken, N.J.: John Wiley & Sons.

Renslow, M., P. Greenfield, and T. Guay. 2000. Evaluation of multi-return LIDAR for forestry applications. *Project Report for the Inventory & Monitoring Steering Committee*, RSAC-2060/4810-LSP-0001-RPT1.

Sabins, F. F. 1997. *Remote Sensing Principles and Interpretation*, 3rd ed. New York: W.H. Freeman and Company.

Short, N. M., Sr. 2009. *The Remote Sensing Tutorial*, http://rst.gsfc.nasa.gov/. Accessed January 25, 2011.

CHAPTER 7

Earth Observation Satellites

7.1 Introduction

In Chap. 1, we discussed that in order for a remote sensor to collect and record energy reflected or emitted from a target or surface, it must reside on a stable platform far away from the target or the surface being observed. Platforms for remote sensors may be situated on the ground (such as a ladder, tall building, or crane), on an aircraft or a balloon within the Earth's atmosphere, or on a spacecraft or satellite outside of the Earth's atmosphere. In Chaps. 4 and 5, we focused our discussion on remote sensing by using aircrafts as the platform—airborne remote sensing. In this chapter, we will learn about spaceborne remote sensing, which mounts sensors on a spacecraft (space shuttle or satellite) orbiting the Earth.

Compared with airborne remote sensing, spaceborne remote sensing provides the following advantages: (1) global coverage; (2) frequent and repetitive observation of an area of interest; (3) immediate transmission of data; (4) digital format of image data ready for computer processing and analysis; and (5) relatively lower cost of image acquisition. We here focus our discussions on satellite remote sensing because satellites provide a more common platform than space shuttles. In particular, we discuss in detail satellite orbital characteristics, major sensing systems employed in satellite remote sensing, and the methods in which remotely sensed data are collected. This chapter ends with introduction of some well-known satellite programs for Earth's resources, meteorological and oceanographic applications, and commercial satellites.

7.2 Satellite Orbital Characteristics

Rockets launch both satellites and space shuttles. The path followed by a satellite is referred to as its orbit. A satellite's orbit is generally elliptical around the Earth. The time taken to complete one revolution of the orbit is called the orbital period. While a satellite revolves around the Earth, the Earth itself is rotating. Therefore, the satellite traces out a different path on the Earth's surface in each subsequent cycle. The sensor onboard the satellite thus "sees" (i.e., takes images) only a certain portion of the Earth's surface along the trail. The area imaged on the surface is referred to as the swath. Satellite sensors usually have different widths of imaging swath, varying from tens to hundreds of kilometers (Fig. 7.1).

In addition to orbital period, a satellite orbit is characterized further by its altitude (the height above the Earth's surface), orientation, and rotation relative to the Earth. Depending on the objective of a satellite mission, a satellite is launched either into a geostationary or near-polar orbit. If a satellite follows an orbit parallel to the equator in the same direction as the Earth's rotation with an orbital period of 24 hours, the satellite would appear stationary with respect to the Earth's surface. This type of orbit is called a geostationary orbit, with which a satellite revolves at the speed that matches the rotation of the

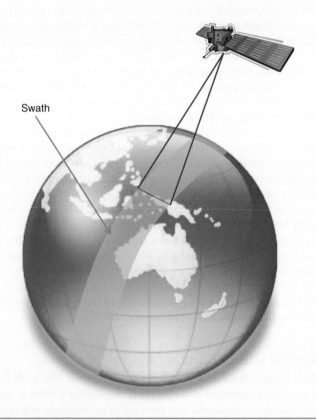

Swath

Figure 7.1   An illustration of imaging swath.

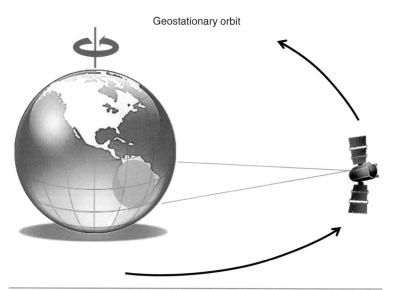

Geostationary orbit

FIGURE **7.2** An illustration of geostationary orbit.

Earth (Fig. 7.2). Geostationary satellites, at an altitude of approximately 36,000 km, can observe and collect data continuously over a specific area at all times. When a series of satellites are launched to image fixed portions (e.g., specific longitudes) of the Earth's surface, they form a constellation of Earth-observing satellites. Such a constellation allows imaging of the whole globe, and can be engineered to image the same region more than one time in a day (24 hours). Geostationary satellites usually provide images of high temporal resolution, but with low spatial resolution. Weather and communications satellites commonly adopt geostationary orbits. Several countries have launched geostationary meteorological satellites to monitor weather and cloud patterns covering specific regions or the entire hemisphere of the Earth. Table 7.1 provides a few examples of geostationary meteorological satellites along with the information about the region under monitoring, launch date, and the operator of each satellite. In Sec. 7.6, we discuss selected meteorological satellites in more detail.

Most Earth resources satellites are devised to follow a near-polar orbit. The orbital plane is inclined at a small angle with respect to the Earth's rotation axis, so that the satellite passes close to the poles and is able to cover nearly the whole surface of the Earth in a repeat cycle (Fig. 7.3). A particular type of near-polar orbit is called a sun-synchronous orbit. The satellite crosses each area of the world at a nearly fixed local time of day (i.e., local solar time) on the sunlit side of the Earth. The design of a sun-synchronous orbit ensures that at any specific latitude, the position of the sun in the sky as the satellite

Satellite	Region Imaged	Longitudes	Launch Date	Operator	Satellite Picture
GOES (Geostationary Operational Environmental Satellite)	East and west coasts of America	GOES-E: 75°W, GOES-W: 135°W	04/1997 (GOES-W), 07/2001 (GOES-E)	United States National Oceanic and Atmospheric Administration	
Meteosat	Europe, Africa, and the Atlantic Ocean	Meteosat 8: 0°, Meteosat 5: 63°E	03/1991 (Meteosat 5), 08/2002 (Meteosat 8 MSG-1)	European Space Agency	
Feng-Yun-2	China and neighboring Asia region	105°E	Jun 25, 2000	China National Satellite Meteorological Center	

MTSAT	Japan and near by Asian and Pacific region	145°E	02/2005 (MTSAT-1R), 02/2006 (MTSAT-2)	Japanese Meteorological Agency	
KALPANA	India, Indian Ocean, and nearby region	74°E	09/2002	India	

TABLE 7.1 Examples of Geostationary Meteorological Satellites

143

Near Polar Orbits

FIGURE 7.3   An illustration of near-polar orbit.

passes overhead will be the same within the same season. The ultimate goal of this type of design is to provide consistent illumination conditions for imaging in a specific region over a series of days, or in a specific season over consecutive years. The images acquired in this way are easier to compare between different times and to mosaic adjacent images together when necessary. Compared with geostationary orbits, near-polar orbits are much lower, typically in the range of 700 to 900 km.

7.3   Earth Resources Satellites

7.3.1   The Landsat Satellite Program

Following the success of early meteorological satellites and manned spacecraft missions, the first unmanned satellite specifically dedicated to multispectral remote sensing entered the planning stages in the late 1960s. The first Landsat satellite, initially called the Earth Resource Technology Satellite (ERTS-1), was carefully designed and constructed by NASA and launched on July 23, 1972. Prior to the launch of the second satellite in the series, the ERTS program was re-named as the Landsat program. All subsequent satellites carried the same designation of Landsat. To date, seven Landsat satellites have been launched. Except for Landsat-6, which failed shortly after the launch, all other satellites were successfully put into orbit. Table 7.2 lists basic characteristics of each of the Landsat missions. All Landsat satellites are placed in near-polar, sun-synchronous orbits. The first three satellites (Landsat-1 to -3) are at an altitude of approximately 900 km and have revisit periods of 18 days, while the later satellites (Landsat-4 to -7) have an orbital height of approximately 700 km and revisit periods of 16 days. All Landsat satellites are designed to cross the equator in the morning (between 9:30 and 10:00 AM) to take advantage of clear skies. A number of sensors have been onboard the Landsat series of satellites, including the Return Beam Vidicon (RBV) camera systems, the MultiSpectral Scanner (MSS) systems, the Thematic Mapper (TM), the Enhanced Thematic Mapper (ETM), and the Enhance Thematic Mapper Plus (ETM+). However, each Landsat mission has a different combination of sensors. Table 7.3 provides a summary of spectral sensitivity and spatial resolution for each sensor. Each of these sensors collected data over a swath width of 185 km, and a full image scene has dimensions of 185 km × 185 km.

The first three Landsat satellites had similar sensors onboard, including RBV and the MSS. The RBV consists of three TV-like cameras, which use color filters to provide multispectral images in the blue-green, yellow-red, and red-IR bands. This sensor failed early on the ERTS-1 and never came into routine use, although it flew again on

Satellite	Launch Date	Decommission Date	Sensors	Orbit Height (km)	Temporal Resolution (days)
Landsat-1	July 23, 1972	January 6, 1978	RBV, MSS	900	18
Landsat-2	January 22, 1975	February 25, 1982	RBV, MSS	900	18
Landsat-3	March 5, 1978	March 31, 1983	RBV, MSS[a]	900	18
Landsat-4	July 16, 1982	June 15, 2001	MSS, TM	705	16
Landsat-5	March 1, 1984	—	MSS, TM	705	16
Landsat-6	Octobar 5, 1993	Failure upon launch	ETM	705	16
Landsat-7	April 15, 1999[b]	—	ETM+	705	16

[a]Band 8 (10.4–12.6 μm) failed after launch.
[b]Scan Line Corrector permanently malfunctioned since May 31, 2003.

TABLE 7.2 Characteristics of Landsat Missions

Sensor	Satellite	Band Width (μm)	Spatial Resolution (m)
RBV	Landsat-1, -2	0.475–0.575	80
		0.580–0.680	80
		0.690–0.830	80
	Landsat-3	0.505–9.750	30
MSS	Landat-1 to -5	0.50–0.60	79/82[a]
		0.60–0.70	79/82
		0.70–0.80	79/82
		0.80–0.11	79/82
	Landsat-3	10.4–12.6[b]	240
TM	Landsat-4, -5	0.45–0.52	30
		0.52–0.60	30
		0.63–0.69	30
		0.76–0.90	30
		1.55–1.75	30
		10.4–12.5	120
		2.08–2.35	30
ETM	Landsat-6	Same as TM	Same as TM
		0.50–0.90	15
ETM+	Landsat-7	Same as TM	30 (60 m thermal)
		0.50–0.90	15

[a]Landsat-1 to -3: 79 m; Landsat-4 to -5: 82 m.
[b]The thermal-IR band of Landsat-3 failed shortly after launch.

TABLE 7.3   Sensors onboard Landsat Missions

Landsat-2. Landsat-3 carried a four-camera RBV array, with each being a panchromatic (0.505–0.750 μm) imager, which provided four contiguous images at 30-m resolution. Figure 7.4 shows examples of precious RBV images, one being the red-IR band image covering the northern New Jersey–New York City region by Landsat-2 and the other a Landsat-3 RBV scene of Cape Canaveral, Florida. The RBV sensor was not included on Landsat missions after Landsat-3 because it appears to be a redundant relative to the MSS.

The Landsat MSS sensor, onboard Landsat-1 to -5, is one of the older generation sensors. The routine data acquisition for MSS was terminated in late 1992, while the MSS onboard Landsat-5 powered off in August 1995. The MSS sensor measures the electromagnetic radiation in four spectral bands from the visible green, red, to the

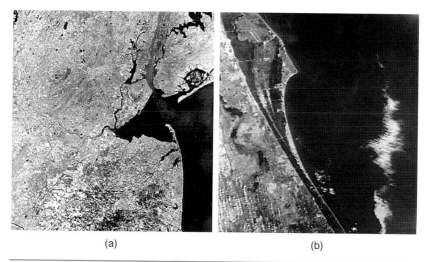

(a)   (b)

FIGURE **7.4**   A Landsat-2 RBV red-IR band image covering the northern New Jersey–New York City region (a) and a Landsat-3 RBV scene of Cape Canaveral, Florida (b).

near-IR wavelengths. It should be noted that the MSS sensor onboard Landsat-3 carried a fifth band in the thermal-IR, but it failed shortly after launch. Each MSS band has a nominal spatial resolution of approximately 79 × 57 m, and a radiometric resolution of 6 bits (64 digital numbers, 0 to 63). Although the MSS sensor has now been replaced by other, more advanced, sensors, the concepts and methods to develop the sensing system and the techniques to interpret satellite images are still used today (Campbell, 2007).

Starting with Landsat-4, the satellite orbit was lowered from 900 km to 705 km, allowing improvement of spatial resolution. The TM sensors, along with MSS, were deployed. However, the TM sensor provided a higher image resolution, sharper spectral separation, and greater radiometric resolution over the MSS sensor. The TM sensors have 30 × 30 m spatial resolution for the visible, near-IR, and mid-IR wavelengths, and 120 × 120 m for the thermal-IR band. Seven, as opposed to four, spectral bands are imaged, with an increase in the number of detectors per band (16 for the non-thermal channels vs. six for MSS). Sixteen scan lines are captured simultaneously for each non-thermal spectral band (four for thermal band), using an oscillating mirror that scans during both the forward (west to east) and reverse (east to west) sweeps of the scanning mirror. The MSS does not acquire data during the return swing (east to west). This difference in scanning mechanism lends the TM sensor a longer dwell time, improving the geometric fidelity and radiometric accuracy of the TM data. The TM sensor was improved to become the ETM for the ill-fated Landsat-6. After the failure of Landsat-6, an improved version of the ETM sensor, named the ETM+, was deployed onboard Landsat-7. The ETM+ maintained similar characteristics as TM, but had an addition panchromatic band of 15-m spatial resolution, and an enhanced thermal-IR band of 60-m resolution. Unfortunately, the scan-line-corrector onboard Landsat-7 permanently malfunctioned after May 31, 2003, which caused a loss of approximately 25% of the data. Most of the data loss was located between scan lines toward the scene edges. Although some gap-filling remedy methods can recover some of the data lost, the gap-filled data cannot meet with the quality of the original data. The longest serving satellite to date is Landsat-5. It was launched in 1984 and remains operational as of the time of this writing (2011), excepting a few temporary technical glitches.

The Landsat program is one of the most successful remote sensing programs in the world. Landsat data have been used by government, commercial, industrial, civilian, and educational communities to support a wide range of applications in such areas as global change research, agriculture, forestry, geology, resource management, geography, mapping, hydrology, and oceanography. The images can be used to map anthropogenic and natural changes on the Earth over a period of several months to two decades. The types of changes that can be identified include agricultural development, deforestation, desertification, natural disasters, urbanization, and the degradation of water resources. The image pair in Fig. 7.5 shows lakes dotting the tundra in northern Siberia to the east of the Taz River (bottom left). These images use near-IR, red, and green wavelength data from the Landsat MSS sensor (1973) and the Landsat ETM+ sensor (2002) to generate color composites. The tundra vegetation is colored a faded red, while lakes appear blue or blue-green. White arrows point to lakes that have disappeared or shrunk considerably between 1973 (top) and 2002 (bottom). After studying satellite imagery of approximately 10,000 large lakes in a 500,000-km^2 area in northern Siberia, scientists documented a decline of 11% in the number of lakes, with at least 125 disappearing completely, because of Arctic warming (Smith, Sheng, MacDonald, and Hinzman, 2005). Another wide application area of Landsat imaegry is urbanization. Figure 7.6 shows the urbanization process in the city of Dubai. Three Landsat images are compared, a Landsat ETM+ image of October 11, 2006, a Landsat-4 TM image of August 28, 1990, and a Landsat-1 MSS image of January 24, 1973. It is seen that between 1973 and 2006, dramatic changes along the coast of the United Arab Emirates followed the development of Dubai, one of the country's seven emirates. The country is located along the eastern coast of the Arabian Peninsula where the land tapers to a sharp tip that nearly separates the Persian Gulf to the north from the Gulf of Oman to the south. The city of Dubai is now home to more than 1.2 million people, and it is still growing rapidly. The city's emergence as a major metropolis and tourist destination is evident in these images.

Looking into the future, the Landsat Data Continuity Mission (LDCM, http://ldcm.nasa.gov), a collaborated effort between NASA

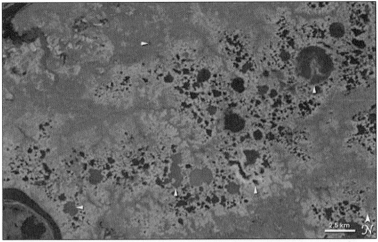

June 27, 1973

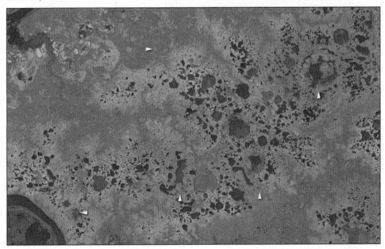

July 2, 2002

FIGURE 7.5 Disappearing Lakes in Siberia. On the top is a Landsat-1 MSS image of June 27, 1973, and at the bottom is a Landsat-7 ETM+ image taken on July 2, 2002. (*Images created by Jesse Allen, Earth Observatory, using data obtained from the University of Maryland's Global Land Cover Facility.*)

and the U.S. Geological Survey (USGS), will provide continuity with the 40-year Landsat land imaging data set. With an earliest projected date of launch in 2012, the LDCM image data will continue to serve widespread routine use in land use planning and monitoring on regional to local scales, disaster response, and water use monitoring. Other uses are also planned, including climate, carbon cycle, ecosystems, water cycle, biogeochemistry, and the Earth's surface/interior.

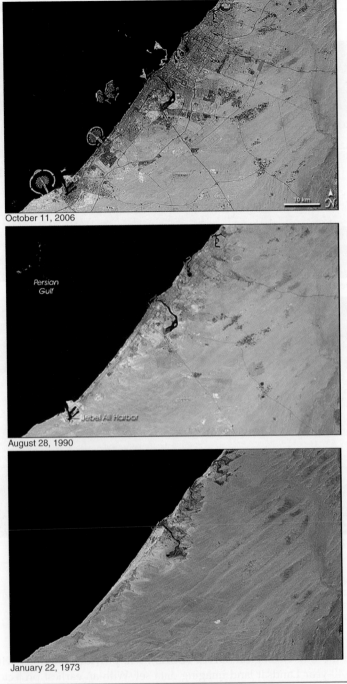

October 11, 2006

Persian
Gulf

Jebel Ali Harbor

August 28, 1990

January 22, 1973

FIGURE 7.6 Urbanization process in the city of Dubai as seen from space. Three Landsat images are compared: a Landsat ETM+ image of October 11, 2006, a Landsat-4 TM image of August 28, 1990, and a Landsat-1 MSS image of January 22, 1973. (*NASA images created by Jesse Allen, Earth Observatory, using data provided by Laura Rocchio, Landsat Project Science Office.*)

The LDCM satellite payload consists of two science instruments: the operational land imager (OLI) and the thermal infrared sensor (TIRS). These two sensors will provide seasonal coverage of the global land-mass at a spatial resolution of 30 m (visible, near-IR, and short wave IR); 100 m (thermal IR); and 15 m (panchromatic). LDCM includes evolutionary advances in technology and performance. The OLI provides two new spectral bands, one tailored especially for detecting cirrus clouds and the other for coastal zone observations, while the TIRS will collect data for two more narrow spectral bands in the thermal region formerly covered by one wide spectral band on Landsat-4 to -7.

7.3.2 SPOT

SPOT (Système Pour l'Observation de la Terre) is a series of Earth observation imaging satellites designed and launched by French Space Agency, CNES (Centre National d'Études Spatiales), with support from Sweden and Belgium. SPOT-1 was launched in 1986, with successors following every three or four years. SPOT-2 was launched on January 22, 1990 and is still operational. SPOT-3 was launched on September 26, 1993, and after 3 years in orbit the satellite stopped functioning. SPOT-4 was launched on March 24, 1998, whereas SPOT-5 was successfully placed into orbit on May 4, 2002. Four SPOT satellites (SPOT-1, -2, -4, and -5) are currently operational. All satellites are in sun-synchronous, near-polar orbits at altitudes of approximately 830 km above the Earth, which results in orbit repetition every 26 days. They have equator-crossing times at approximately 10:30 AM local solar time. SPOT was designed to be a commercial provider of Earth observation data, and provides nearly global coverage, between 87°N and 87°S.

SPOT satellites use along-track scanning technology. Each satellite carries two high-resolution visible (HRV) imaging systems, which can be operated independently and simultaneously. The HRV sensors are capable of sensing in either single-channel panchromatic (PLA) mode or three-channel multispectral (MLA) mode. Each sensor consists of four linear arrays of detectors: one 6000-element array for the PLA mode recording at a spatial resolution of 10 m and three 3000-element arrays for each of the three MLA bands, recording at 20 m spatial resolution. The swath width for both modes is 60 km at nadir. SPOT-5 improves the ground resolution to 5 and 2.5 m in PLA mode and 10 m in MLA mode in all three spectral bands in the visible and near-IR ranges. Table 7.4 highlights the spectral and spatial characteristics of the HRV sensors onboard SPOT-1 to -5.

Another innovation with the SPOT satellite program is the off-nadir viewing capacity of the sensors. The ability to point the sensors up to 27° from the nadir to look to either side of the satellite's vertical track allows SPOT to view within a 950-km swath and to revisit any location several times per week, thereby increasing the satellite's revisit capability. For example, this oblique viewing capability

Sensor	Satellite	Band	Wavelength Range (μm)	Spatial Resolution (m)
HRV	SPOT-1, -2, -3	XS1	0.50–0.59	20
		XS2	0.61–0.68	20
		XS3	0.79–0.89	20
		PAN	0.51–0.73	10
HRVIR	SPOT-4	XI1	0.50–0.59	20
		XI2	0.61–0.68	20
		XI3	0.79–0.89	20
		XI4	1.53–1.75	20
		M	0.61–0.68	10
HRG	SPOT-5	B1	0.50–0.59	10
		B2	0.61–0.68	10
		B3	0.79–0.89	10
		SWIR	1.58–1.75	20
		PAN	0.51–0.73	5 (or 2.5 m in supermode)[a]
HRS	SPOT-5	PAN	0.51–0.73	10[b]
VEGETATION	SPOT-4, -5	B0	0.43–0.47	1000
		B2	0.61–0.68	1000
		B3	0.79–0.89	1000
		MIR	1.58–1.75	1000

[a]HRG panchromatic band has two linear arrays with 5-m spatial resolution that can be combined to produce 2.5-m resolution images.
[b]HRS panchromatic band has a spatial resolution of 10 m, but has a 5-m sampling along the track.

TABLE 7.4   Sensors onboard SPOT Satellites

increases the revisit frequency of equatorial regions to three days (seven times during the 26-day orbital cycle). Areas at the latitude of 45° can be imaged more frequently (11 times in 26 days) due to the convergence of orbit paths toward the poles. As the sensors point away from the nadir, the swath varies from 60 to 80 km in width. By pointing both HRV sensors to cover adjacent ground swaths at the nadir, a swath of 117 km (3-km overlap between the two swaths) can be imaged. In this mode of operation, either PLA or MLA data can be collected, but not both simultaneously (Canada Centre for Remote Sensing, 2007). This off-nadir viewing technology not only improves the ability to monitor specific locations and increases the chances of obtaining cloud free scenes, but also provides the capability of acquiring images for stereoscopic viewing. As we discussed in Chap. 5, stereo pair imagery is vital for 3D terrain modeling and visualization, taking height measurements, and topographic mapping.

SPOT's fine spatial resolution and pointable sensors have gained its popularity among a variety of user communities. The applications of SPOT data are numerous, ranging from urban and rural planning, to environmental monitoring, natural disaster management, and oil and gas exploration. SPOT allows applications requiring both fine spatial details and short revisit, such as monitoring urban environments and the spread of infectious diseases. The acquisition of stereoscopic imagery from SPOT has played an important role in mapping applications and in the derivation of topographic information (DEMs) from satellite data. Applications requiring frequent monitoring, such as agriculture and forestry, can be well served by the SPOT sensors as well. Figure 7.7 show an example of a SPOT-2 HRV multispectral

FIGURE 7.7   False color composite image of Los Angeles, California, U.S.A., which is created using SPOT-2 HRV multispectral bands on May 27, 1998. Urban and built-up areas appear bluish, while the Pacific Ocean is dark blue. Vegetation is shown in red. We can also clearly see clouds scattered over the metropolitan area.

Figure 7.8   SPOT-2 image of Lake Okeechobee, Florida, U.S.A., taken by an HRV sensor on April 3, 1998. Healthy vegetation appears in red, while the lake itself is dark blue. Apparently, a good amount of clouds is shadowed over the land to the east and south. Lake Okeechobee, locally referred to as The Lake or The Big O, is the largest freshwater lake in the state of Florida. Lake Okeechobee covers 730 square miles (1890 km²), approximately half the size of the state of Rhode Island. The floor of the lake is a limestone basin. Lake Okeechobee is exceptionally shallow for a lake of its size, with an average depth of only 9 feet (3 m). The 100-foot (30-m) wide dike surrounding the lake is part of the Florida Trail, a 1400-mile long trail that is a National Scenic Trail.

image that covers the city of Los Angeles, California, U.S.A. Figure 7.8 is a false color composite of a SPOT-2 HRV MLA image centering at Lake Okeechobee, Florida, U.S.A., the second largest freshwater lake within the lower 48 states. It is especially deserved to note that both SPOT-4 and -5 satellites carry onboard a low-resolution wide-coverage instrument for monitoring the Earth's vegetated cover. The VEGETATION instrument provides global coverage on a nearly daily basis at a resolution of 1 km with a swath of 2250 km, enabling the observation of long-term regional and global environmental changes. This sensor images four spectral bands in blue, red, near-IR, and mid-IR wavelengths. Unlike other sensors onboard SPOTs, the VEGETATION program is co-financed by the European Union, Belgium, France, Italy, and Sweden.

7.3.3 Indian Remote Sensing (IRS) Satellites

India has successfully launched several Earth-resources satellites that gather data in the range from visible, through near-IR to mid-IR bands, beginning with IRS-1A on March 17, 1988. IRS-1 is India's dedicated Earth resources satellite system operated by Indian Space Research Organization (ISRO) and the National Remote Sensing Agency. IRS-1B, launched on August 29, 1991, has similar orbital characteristics as IRS-1A. These satellites are run on a sun-synchronous orbit with an orbital height of 904 km and 22 days revisit at the equator. Commencing with the third satellite in the series, IRS-1C (launched December 28, 1995), the orbital height was lowered to 817 km, but the revisit time was prolonged to 24 days. RESOURCESAT-1 is the tenth satellite of ISRO in the IRS series, intended to not only continue the remote sensing data services provided by IRS-1C and IRS-1D (launched on September 27, 1997), both of which have far outlived their designed mission life periods, but also to vastly enhance the data quality. RESOURCESAT-1 is the most advanced remote sensing satellite built by ISRO as of 2003.

The IRS satellite sensors combine features from both the Landsat MSS/TM sensors and the SPOT HRV sensor (Canada Centre for Remote Sensing, 2007). The third and fourth satellites in the series, IRS-1C and IRS-1D, both carried three sensors: a single-channel panchromatic (PAN) high-resolution camera (5.8 m), a medium-resolution four-channel linear imaging self-scanning sensor (LISS-III, 23 m for bands 2–4 and 50 m for band 5), and a coarse resolution two-channel wide field sensor (WiFS, 188 m). Table 7.5 outlines the specific characteristics of each sensor. RESOURCESAT-1 carried an enhanced version of LISS, LISS-IV, which has spatial resolution of 5.8 m in the visible and near-IR bands (i.e., bands 2–4), as well as an advanced wide field sensor (AWiFS), which improves its spectral resolution to four bands and the spatial resolution to 56–70 m. Like the French SPOT sensor, the panchromatic sensor of IRS can be steered up to 26° across-track, enabling stereoscopic imaging and increased revisit capabilities to as few as five days. The high-resolution panchromatic and LISS-IV multispectral data are useful for urban and mapping applications, detailed vegetation discrimination, and land-cover mapping. The WiFS and AWiFS sensors balance the spatial resolution and coverage, and provide an excellent data source for regional scale vegetation monitoring, land cover, and resources planning. Figure 7.9 shows some sample images of LISS-III panchromatic and multispectral bands and WiFS.

7.3.4 MODIS

The moderate resolution imaging spectroradiometer (MODIS) is a key instrument designed to be part of NASA's EOS to provide long-term observation of the Earth's land, ocean, and atmospheric properties and dynamics. The MODIS sensor was developed upon the heritage

Sensor	Satellite	Bands	Bandwidths (µm)	Resolution (m)	Swath Width (km)
Linear Imaging Self-Scanning System I (LISS-I)	IRS-1A, 1B	LISS-I-1 LISS-I-2 LISS-I-3 LISS-I-4	0.45–0.52 (blue) 0.52–0.59 (green) 0.62–0.68 (red) 0.77–0.86 (near IR)	72	148
Linear Imaging Self-Scanning System II (LISS-II)	IRS-1A, 1B	LISS-II-1 LISS-II-2 LISS-II-3 LISS-II-4	0.45–0.52 (blue) 0.52–0.59 (green) 0.62–0.68 (red) 0.77–0.86 (near IR)	36	74
Linear Imaging Self-Scanning System III (LISS-III)	IRS-1C, 1D, RESOURCESAT-1	LISS-III-2 LISS-III-3 LISS-III-4 LISS-III-5	0.52–0.59 (green) 0.62–0.68 (red) 0.77–0.86 (near IR) 1.55–1.70 (mid-IR)	23 23 23 50	142 142 142 148
		PAN	0.5–0.75	5.8	70
High Resolution Linear Imaging Self-Scanning System IV (LISS-IV)	RESOURCESAT-1	LISS-IV-2 LISS-IV-3 LISS-IV-4	0.52–0.59 (green) 0.62–0.68 (red) 0.77–0.86 (near IR)	5.8	24–70
Wide Field Sensor (WiFS)	IRS-1C, 1D	WiFS-1 WiFS-2	0.62–0.68 (red) 0.77–0.86 (near IR)	188	774
Advanced Wide Field Sensor (AWiFS)	RESOURCESAT-1	AWiFS-1 AWiFS-2 AWiFS-3 AWiFS-4	0.52–0.59 (green) 0.62–0.68 (red) 0.77–0.86 (near IR) 1.55–1.70 (mid-IR)	56–70	370–740

TABLE 7.5 Sensors onboard Indian Remote Sensing (IRS) Satellites

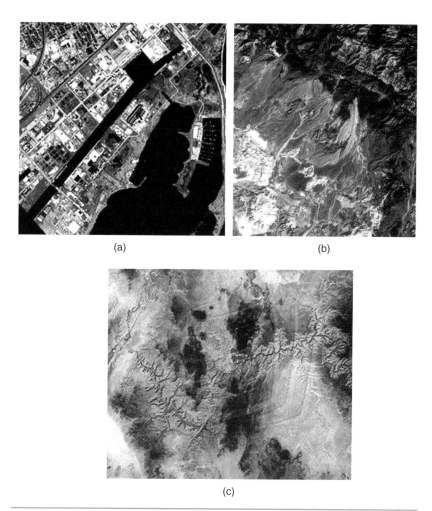

(a)   (b)

(c)

FIGURE 7.9   Sample images of IRS sensors. (a) LISS-III panchromatic image of Toronto, Canada (5-m resolution); (b) a three-band color composite made from the 23-m LISS-III, showing mountainous terrain and pediments with alluvium fans in southern Iran; and (c) a WiFS multispectral image (188-m resolution) of the Grand Canyon, Arizona. (Image (a) *was acquired from the website of Professor Joe Piwowar, Department of Geography, University of Regina, Saskatchewan, Canada. Images (b) and (c) were acquired through the "Remote Sensing Tutorial" created and maintained by Dr. Nicholas M. Short.*)

of several earlier sensors, such as the advanced very-high-resolution radiometer (AVHRR), coastal zone color scanner (CZCS), and the Landsat TM. Therefore, the design of MODIS aims at providing not only a continuous global observation, but also a new generation of sensor with improved spectral, spatial, radiometric, and temporal resolutions. In addition to advances in the sensor design, the MODIS mission also emphasizes the development of operational data-processing algorithms, validated data products, and models capable of advancing the understanding of global change and of assisting policy makers in making sound decisions concerning the protection of our environment. Two MODIS instruments onboard NASA's Terra and Aqua satellites, respectively, were successfully put in orbit on December 18, 1999 (EOS-Terra) and May 4, 2002 (EOS-Aqua).

EOS-Terra and EOS-Aqua are polar-orbiting, sun-synchronous, near-polar, and circular platforms. The orbit height of EOS platforms is 705 km at the equator. Terra crosses the equator (descending mode) at approximately 10:30 AM local time, while Aqua crosses (ascending) the equator at approximately 1:30 PM. Each MODIS instrument has a two-sided scan mirror that scans the Earth's surface perpendicularly to the spacecraft track. The mirror scanning can extend to 55° at either side of the nadir, providing a swath of 2330 km. The wide swath allows each instrument to scan the whole globe every one to two days. The design life of each sensor is six years.

In addition to the high temporal resolution, the MODIS sensor has high spectral, spatial, and radiometric resolutions compared to previous sensor systems, such as the AVHRR. A total of 36 spectral bands were carefully positioned across the 0.412 to 14.235 μm spectral region (Table 7.6). Among the 36 spectral bands, the first two bands in the red and near-IR regions have a spatial resolution of 250 m. There are five additional bands with a spatial resolution of 500 m located in the visible to short-wave IR (SWIR) spectral regions. The remaining 29 spectral bands (bands 8 to 36) have 1000-m spatial resolution, and are located in the middle and longwave thermal IR (TIR) regions. The MODIS instrument also has a 12-bit radiometric resolution and an advanced onboard calibration subsystem that ensures high calibration accuracy (Justice et al., 1998). MODIS has proven to be a well-suited sensor for a wide range of applications centering at the understanding of land, ocean, and atmospheric processes, and the effects of human activities on the global environment. Figure 7.10 shows a monthly average of global normalized difference vegetation index (NDVI, a measurement of vegetation amount), based on observations from the MODIS on NASA's Terra satellite between November 1 and December 1, 2007. Satellites can be used to monitor how "green" different parts of the planet are and how the greenness changes over time. The greenness values on a map range from –0.1 to 0.9. Higher values (dark greens) show land areas with plenty of leafy green vegetation, such as the Amazon Rainforest. Lower values (beige to white) show areas with little or no vegetation, including sand seas and Arctic areas. Areas with moderate amounts of vegetation are pale green. Land areas with no data appear gray, and water appears blue. These observations help scientists understand the influence of natural cycles, such as drought and pest outbreaks, on vegetation, as well as human influences, such as land-clearing and global warming. Figure 7.11 is a false color image of global sea

Primary Use	Band	Band Width (μm)	Spatial Resolution (m)
Land/Cloud Boundaries	1	0.620–0.670	
	2	0.841–0.876	250 m
Land/Cloud Properties	3	0.459–0.479	
	4	0.545–0.565	
	5	1.230–1.250	
	6	1.628–1.652	
	7	2.105–2.155	500 m
Ocean Color/	8	0.405–0.420	
Phytoplankton/	9	0.438–0.448	
Biogeochemistry	10	0.483–0.493	
	11	0.526–0.536	
	12	0.546–0.556	
	13	0.662–0.672	
	14	0.673–0.683	
	15	0.743–0.753	
	16	0.862–0.877	1000 m
Atmospheric Water Vapor	17	0.890–0.920	
	18	0.931–0.941	
	19	0.915–0.965	1000 m
Surface/Cloud Temperature	20	3.660–3.840	
	21	3.929–3.989	
	22	3.929–3.989	
	23	4.020–4.080	1000 m
Atmospheric Temperature	24	4.433–4.598	
	25	4.482–4.549	1000 m
Cirrus Clouds	26	1.360–1.390	1000 m
Water Vapor	27	6.535–6.895	
	28	7.175–7.475	
	29	8.400–8.700	1000 m
Ozone	30	9.580–9.880	1000 m
Surface/Cloud Temperature	31	10.780–11.280	
	32	11.770–12.270	1000 m
Cloud Top Altitude	33	13.185–13.485	
	34	13.485–13.785	
	35	13.785–14.085	
	36	14.085–14.385	1000 m

Note: Bands 21 and 22 are similar, but Band 21 saturates at 500K and Band 22 at 328K.

TABLE 7.6   MODIS Spectral Bands and Primary Uses

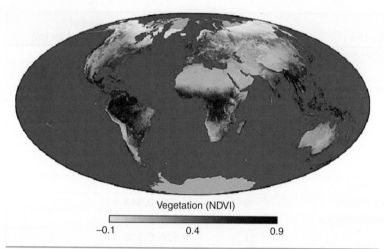

Vegetation (NDVI)

−0.1   0.4   0.9

FIGURE 7.10   Monthly average of the global Normalized Difference Vegetation Index (NDVI), based on observations from the MODIS on NASA's Terra satellite between November 1 and December 1, 2007. (*Courtesy of NASA's Earth Observatory. NASA images produced by Reto Stockli and Jesse Allen, using data provided by the MODIS Land Science Team.*)

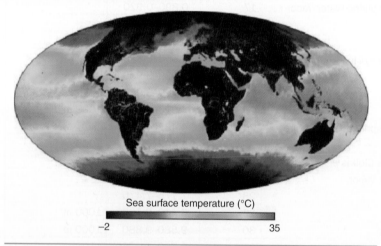

Sea surface temperature (°C)

−2   35

FIGURE 7.11   Global sea surface temperature by Terra's MODIS. Every day the MODIS measures sea surface temperature over the entire globe with high accuracy. This false-color image shows a one-month composite for May 2001. (*Image courtesy MODIS Ocean Group, NASA GSFC, and the University of Miami; and text after NASA's Earth Observatory.*)

surface temperature showing a one-month composite for May 2001. Red and yellow indicate warmer temperatures, green is an intermediate value, while blues and purples are progressively colder values. The MODIS sea surface temperature maps are particularly useful

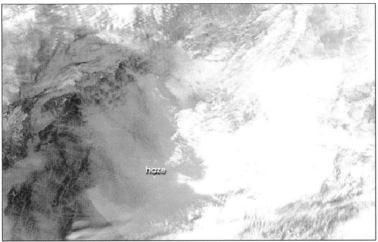

December 28

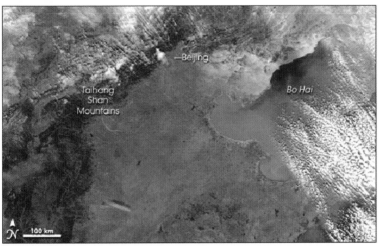

December 31

FIGURE 7.12 Aqua's MODIS images of Beijing and the nearby region on December 28, 2007 (top) and December 31, 2007 (bottom). In the top image, a combination of haze and clouds almost completely obscures the land surface, leaving only parts of the Taihang Mountains visible. In the bottom image, sparse streaks of clouds dot the otherwise clear sky. Snow dusts the ground east and northeast of the city, and Bo-Hai is thick with tan-colored sediment. The northerly winds from Siberia often play an important role in Beijing's air quality. (*Image courtesy of the MODIS Rapid Response Team at NASA GSFC.*)

in studies of temperature anomalies, such as El Niño, and how air–sea interactions drive changes in weather and climate patterns. Figure 7.12 shows Aqua's MODIS images of Beijing (the capital of China) and the nearby region on December 28 and December 31, 2007.

Beijing's air quality has been a serious concern of local residents and visitors. The northerly wind can carry a large amount of mineral dust from the Gobi desert, while urban construction over the past two decades also generates a large amount of dust. Coal-burning power plants and factories produce sulfate and soot particles, while a recent dramatic increase of motor vehicles also generates significant amounts of nitrates and other particles. Agricultural burning and diesel engines further yield dark sooty particles that scientists call black carbon. These natural and anthropogenic pollutants have made Beijing one of the most polluted cities in the world. MODIS sensors provide an effective means for timely monitoring of air quality over Beijing and other localities in the world.

7.4   Meteorological and Oceanographic Satellites

7.4.1   NOAA POES Satellites

NOAA-POES satellites are a series of polar-orbiting environmental satellites operated by NOAA. The satellite program began with the TIROS-N satellite in 1978. Some older generations of the POES satellites are out of service today, but there are still quite a few operational satellites in orbit. Table 7.7 includes brief information on the missions starting from NOAA-11 to NOAA-19, the most recent one. NOAA POES satellites are engineered in sun-synchronous, near-polar orbits (830 to 870 km above the Earth), aimed at providing complementary information to the geostationary meteorological satellites (such as GOES). At any time, NOAA ensures at least two active POES satellites are operational in orbit, so that it can provide at least four image acquisitions per day for any location on the Earth. One satellite crosses the equator in the early morning from north-to-south (descending), while the other crosses in the afternoon from south-to-north (ascending).

The AVHRR is the primary sensor onboard the NOAA POES satellites, used for meteorology, and global and regional monitoring. The AVHRR sensor detects radiation in the visible, near-, mid-, and thermal-IR portions of the electromagnetic spectrum with spatial resolution of 1.1 km over a swath width of 3000 km. The earlier AVHRR instrument on NOAA-11 to -14 had five bands in the visible (band 1), near-IR (band 2), mid-IR (band 3) and thermal-IR (bands 4 and 5) regions (Table 7.8). The newer AVHRR sensor starting from NOAA-15 has an extra band in the SWIR (designated band 3A), which shares the same transmission channel with the mid-IR band (now designated band 3B). At any instant in time, only one of the 3A or 3B bands is activated.

Satellite	Launch Date	Decommission Date/ Operational Status	Ascending Time	Descending Time
NOAA-11	09/24/1988	06/16/2004	1340	0140
NOAA-12	05/14/1991	08/10/2007	1930	0730
NOAA-14	12/30/1994	05/23/2007	1340	0140
NOAA-15	05/13/1998	AM Secondary	1930	0730
NOAA-16	09/21/2000	PM Secondary	1400	0200
NOAA-17	06/24/2002	AM Backup	2200	1000
NOAA-18	05/20/2005	PM Secondary	1400	0200
NOAA-19	02/06/2009	PM Primary	1336	

Note: NOAA-13 launched August 9, 1993, failed due to an electrical short circuit in the solar array.
METOP-A provides the data for AM Primary orbit. This mission is owned and operated by EUMETSAT.

TABLE 7.7   Orbital Characteristics of NOAA POES Satellites

Band	Wavelength Range (µm) NOAA 11-14	Wavelength Range (µm) NOAA 15-19	Spectral Region	Spatial Resolution (km)	Applications
1	0.58–0.68	0.58–0.68	Visible	1.1	Vegetation mapping, NDVI, daytime cloud, snow, ice
2	0.725–1.11	0.725–1.11	Near-IR	1.1	Vegetation mapping, NDVI, land/water interface, snow, ice
3	3.55–3.93	3A: 1.58–1.64, 3B: 3.55–3.93	3A: Short-wave-IR, 3B: Mid-IR	1.1	Snow/ice discrimination, day/night cloud and surface temperature mapping
4	10.30–11.30	10.30–11.30	Thermal-IR	1.1	Day/night cloud and surface temperature mapping
5	11.50–12.50	11.50–12.50	Thermal-IR	1.1	Cloud and surface temperature, day/night cloud mapping

TABLE 7.8   AVHRR Spectral Bands and Applications

AVHRR data can be acquired in four operational modes, differing in spatial resolution and the method of transmission. Data can be transmitted directly to the ground and viewed as data are collected, or recorded onboard the satellite for later transmission and processing. High-resolution picture transmission (HRPT) data are full resolution (1.1 km) image data transmitted to a local ground station as they are being collected. Local area coverage (LAC) are also full resolution (1.1 km) data, but recorded with an onboard tape recorder for subsequent transmission during a station overpass. Automatic picture transmission (APT) provides low-resolution (4 km) direct transmission and display. Global area coverage (GAC) data, with spatial resolution of 4 km, provides daily sub-sampled global coverage recorded on tape recorders and then transmits to a ground station. Many weather stations around the world operate ground stations that routinely receive real-time HRPT data. LAC data are available from NOAA's Satellite Active Archive (SAA) on the Internet (http://www.saa.noaa.gov). Since onboard tape facility is limited, only a limited number of scenes are archived. GAC data are also available from the NOAA SAA.

AVHRR image data have been widely used for meteorological studies. Figure 7.13 shows an image of Hurricane Katrina on August 27, 2005 and the nearby Gulf of Mexico region, using AVHRR data of

FIGURE 7.13   Image of Hurricane Katrina on August 27, 2005 and the nearby Gulf of Mexico region, a red-green-blue (RGB) composite using AVHRR data of channels 1, 2, and 4. (*Image courtesy of NOAA CoastWatch—East Coast Node. http:// coastwatch.chesapeakebay.noaa.gov/cw_about.html.*)

channels 1, 2, and 4 to generate an RGB composite. Its short temporal revisit offers a major advantage for tracking fast-changing surface conditions and monitoring regional and global environmental environments and ecosystems. Its application areas range from mapping of snow, ice, and sea surface temperature, to monitoring crop conditions, analyzing vegetation patterns, and studying urban heat islands. AVHRR data have been gaining popularity for global modeling. Figure 7.14 shows an AVHRR image of the West Ross Ice Shelf, Antarctica, on January 27, 2001, created by the National Snow and Ice Data Center (http://nsidc.org/data/iceshelves_images/). In addition, weekly and bi-weekly NDVI composites of the entire United States are archived and available at no charge to the public from January 1989 to the present. These NDVI images are used by USDA policy officials for monitoring vegetation conditions in cropland and pasture regions throughout the growing season. Figure 7.15 displays vegetation condition images for the conterminous United States, based on NOAA-18 NDVI weekly composites. The two images show the difference in vegetated conditions between early May and early June in 2010. The comparison of such images allows us to examine how much vegetated cover changed within a month.

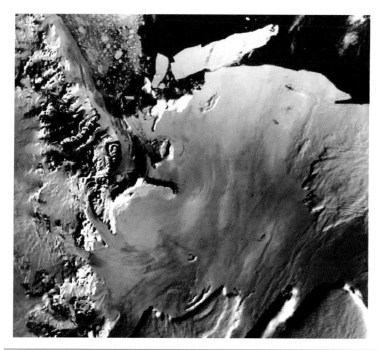

FIGURE 7.14   AVHRR image of West Ross Ice Shelf, Antarctica, on January 27, 2001. (*Image Courtesy of National Snow and Ice Data Center's Images of Antarctic Ice Shelves. http://nsidc.org/data/iceshelves_images/.*)

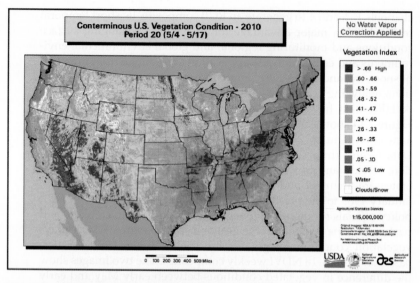

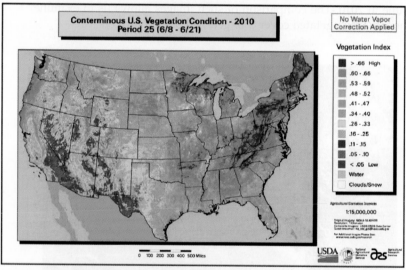

FIGURE 7.15   Vegetation condition images for the conterminous United States were created based on NOAA-18 NDVI weekly composites. (*Images courtesy of the NASS Spatial Analysis Research Section [SARS], USDA. http://www.nass.usda.gov/research/avhrr/avhrrmnu.htm.*)

7.4.2 GOES

The GOES (Geostationary Operational Environmental Satellite) program is a key element in NOAA's National Weather Service (NWS) operations. GOES weather imagery and sounding data provide a continuous and reliable stream of environmental information used to support weather forecasting, severe storm tracking, and other meteorological tasks. Technological improvements in the geostationary satellite system since 1974 have been responsible for making the current GOES system the basic element for U.S. weather monitoring and forecasting. The first weather satellite, TIROS-1 (Television and Infrared Observation Satellite-1), was launched in 1960 by the United States. Several other weather satellites were launched later into near-polar orbits, providing repetitive coverage of global weather patterns. In 1966, NASA launched the geostationary Applications Technology Satellite (ATS-1), which provided hemispheric images of the Earth's surface and cloud cover every half hour. The development and movement of weather systems could then be routinely monitored. The GOES system is a follow-up to the ATS series. For more than 30 years, the GOES series of satellites have been extensively used by meteorologists for weather monitoring and forecasting. These satellites are part of a global network of meteorological satellites spaced at approximately 70° longitude intervals around the Earth to provide near-global coverage. Two GOES satellites, placed in geostationary orbits 36,000 km above the equator, each view approximately one-third of the Earth. One is located at 75°W longitude and monitors North and South America and most of the Atlantic Ocean. The other is situated at 135°W longitude and monitors North America and the Pacific Ocean basin. Together they cover the region from 20°W to 165°E longitude. Two generations of GOES satellites have been launched, each measuring emitted and reflected radiation from which atmospheric temperature, winds, moisture, and cloud cover can be derived. The first generation of satellites consists of GOES-1 (launched in 1975) through GOES-7 (launched in 1992). The sensor onboard these satellites was visible and IR spin-scan radiometer (VISSR). Due to their design, these satellites could image the Earth for only a small percentage of the time (approximately 5%). The second generation of satellites began with GOES-8 (launched in 1994) and has numerous technological improvements over the first series. They provide near-consecutive observation of the Earth, allowing more frequent imaging (as often as every 15 minutes). This increase in temporal resolution coupled with improvements in the spatial and radiometric resolution of the sensors provides timely information and improves data quality for forecasting meteorological conditions. As of the writing of this book, GOES-8, -9, and -10 are all decommissioned, but GOES-11 to -15 are operational. GOES-11 (launched on May 3, 2000) is for monitoring the 135°W region, while GOES-13 (launched on May 24, 2006) is for the 75°E region. GOES-12 is for imaging South America. GEOG-14 and -15 are in the status of on-orbit storage and standby, respectively.

The second-generation GOES satellites carry two instruments, an imager and a sounder. The GOES I-M Imager is a five-channel (one visible, four IR) imaging radiometer designed to sense radiant and

Band	Wavelength Range (μm)	Spectral Region	Spatial Resolution	Application
1	0.55–0.75	Visible	1 km	Cloud, pollution, and haze detection; severe storm identification
2	3.8–4.0	Shortwave IR	4 km	Identification of fog at night; discriminating water clouds and snow or ice clouds during daytime; detecting fires and volcanoes; night time determination of sea surface temperatures
3	6.5–7.0	Moisture	8 km	Estimating regions of mid-level moisture content and advection; tracking mid-level atmospheric motion
4	10.2–11.2	Longwave IR	4 km	Identifying cloud-drift winds, severe storms, and heavy rainfall
5	11.5–12.5	Longwave IR	4 km	Identification of low-level moisture; determination of sea surface temperature; detection of airborne dust and volcanic ash

TABLE 7.9   GOES Imager Spectral Bands

solar reflected energy from sampled areas of the Earth. Table 7.9 outlines individual spectral bands, their spatial resolution, and meteorological applications. The IR bands collect both daytime and nighttime images. By using a mirror scanning system in conjunction with a Cassegrain telescope, the sensor can simultaneously sweep an 8-km north-to-south swath along an east-to-west/west-to-east path, at a rate of 20 degrees (optical) east-west per second. This translates into being able to scan a 3000 × 3000 km (1864 × 1864 mi) "box" centered over the United States in just 41 seconds. In fact, the sensor pointing and scan selection capability enable it to image the entire hemisphere. Small-scale imaging of selected areas is also possible, which allows for monitoring specific weather trouble spots and improving short-term forecasting. The Imager data are in 10-bit radiometric resolution, and can be transmitted directly to local users on the Earth's surface. The Sounder measures emitted radiation in 18 thermal-IR bands and 1 visible band. The Sounder's data have a spatial resolution of 8 km and 13-bit radiometric resolution. Sounder data are used for surface and cloud-top temperatures, multi-level moisture profiling in the atmosphere, and ozone distribution analysis.

Figure 7.16   GOES image showing Hurricane Floyd just off the Florida coast on September 14, 1999. (*Image courtesy of NASA Earth Observatory. Image created by Marit Jentoft-Nilsen, NASA GSFC Visualization Analysis Lab.*)

GOES images are frequently used to monitor weather systems over the southeastern United States and the adjacent ocean areas where many severe storms originate and develop. Figure 7.16 is a GOES image showing Hurricane Floyd approaching the Florida coast on September 14, 1999. With hurricane-force winds extending 125 mi from the storm's eye and sustained winds up to 140 mph, Floyd threatened coastal areas in southeastern states, forcing many evacuations, including NASA's Kennedy Space Center. Hurricane Katrina moved ashore over southeast Louisiana and southern Mississippi early on August 29, 2005, as an extremely dangerous Category-4 storm. With winds of 135 mph (217 km/h), a powerful storm surge, and heavy rains, Katrina pounded the U.S. Gulf Coast, triggering extensive life-threatening flooding. Figure 7.17 is a GOES image showing the storm moved over southern Mississippi at 9:02 AM. The eye of the storm was due east of New Orleans, Louisiana.

7.4.3   Envisat Meris

The **ME**dium **R**esolution **I**maging **S**pectrometer (MERIS) was launched on January 3, 2002 by the European Space Agency (ESA) onboard its polar orbiting Envisat-1 Earth observation satellite. MERIS is primarily

Figure 7.17   GOES image showing Hurricane Katrina moving ashore over southeast Louisiana and southern Mississippi early on August 29, 2005, as an extremely dangerous Category 4 storm. The eye of the storm was due east of New Orleans, Louisiana. (*Images courtesy of GOES Project Science Office.*)

dedicated to ocean color observations, but is also used in atmospheric and land surface related studies. MERIS has a high spectral and radiometric resolution and a dual spatial resolution within a global mission covering open ocean and coastal waters, and a regional mission covering land surfaces.

MERIS is a programmable, medium-spectral resolution imaging spectrometer. Fifteen spectral bands can be selected by ground command, each of which has a programmable width and a programmable location in the spectral range from 390 to 1040 nm. Table 7.10 describes briefly the individual spectral bands and their oceanographic and other applications. The spatial resolution of the sensor is 300 m (at nadir), but this resolution can be reduced to 1200 m by the onboard combination of four adjacent samples across-track over four successive lines. Depending on the sizes of various targets to be observed and the diversity of their spectral and radiometric properties, the spatial, spectral, and radiometric resolutions can be programmed. This programmability allows for selecting width and position of a respective spectral band and tuning the dynamic range, thus making it adaptable to different target observation, which is considered to be a high priority during the MERIS mission.

MDS Nr.	Band Centre (NM)	Bandwidth (NM)	Potential Applications
1	412.5	10	Yellow substance and detrital pigments
2	442.5	10	Chlorophyll absorption maximum
3	490	10	Chlorophyll and other pigments
4	510	10	Suspended sediment, red tides
5	560	10	Chlorophyll absorption minimum
6	620	10	Suspended sediment
7	665	10	Chlorophyll absorption and fluorescence reference
8	681.25	7.5	Chlorophyll fluorescence peak
9	708.75	10	Fluorescence reference, atmospheric corrections
10	753.75	7.5	Vegetation, cloud
11	760.625	3.75	Oxygen absorption R-branch
12	778.75	15	Atmosphere corrections
13	865	20	Vegetation, water vapor reference
14	885	10	Atmosphere corrections
15	900	10	Water vapor, land

TABLE 7.10   ENVISAT MERIS Spectral Bands

MERIS employs the pushbroom scanning technology. CCD arrays provide spatial sampling in the across-track direction, while the satellite's motion provides scanning in the along-track direction. MERIS is designed in such a way that it can acquire data over the Earth's surface whenever illumination conditions are suitable. The instrument's 68.5° field of view around nadir yields an image swath of 1150 km. This wide field of view is shared between five identical optical modules arranged in a fan-shaped configuration. In the calibration mode, correction parameters such as offset and gain are generated, which are then used to correct the recorded spectra. This correction can be carried out either onboard or on the ground.

The main usage of MERIS data is the study of the upper layers of the ocean by the observations of water color. Figure 7.18 shows a MERIS image, taken on September 12, 2003, covering the French territory of New Caledonia and surrounding ocean in the southwest Pacific. New Caledonia comprises a main island (Grande Terre), the Loyalty Islands, and several smaller islands, and has a land area of 18,575.5 km^2 (7172 mi^2). In addition, the MERIS mission contributed to the understanding of atmospheric parameters by providing data on cloud-top height and optical thickness, water vapor column

FIGURE 7.18   MERIS image taken on September 12, 2003 shows the main island of the French territory of New Caledonia, Grande Terre (in the lower part), the Loyalty Islands (to the northeast), and the surrounding ocean. (*Image courtesy of European Space Agency.*)

content, as well as aerosol properties. Figure 7.19 is a MERIS image taken on May 4, 2004 showing a large sand/dust storm from the Sahara Desert making its way into southern Italy. This warm wind from the Sahara is known as the Sirocco in Italy. When the air becomes moist, the sand and dust can then fall with the rain. In terms of land imaging, MERIS focuses on vegetation patterns and other land surface parameters related to global change climate dynamics and biogeochemical cycles. Figure 7.20 shows a MERIS false color image of the East African island of Madagascar on August 13, 2003. The Comoros Islands can also be seen to the northwest just poking through the clouds. The bright yellow in the east and the north indicates tropical rain forest. The more red tones in the middle represent highland areas. To the west of the island, the green tones represent deciduous forest.

7.4.4   Other Satellites

The United States operates the Defense Meteorological Satellite Program (DMSP) series of satellites, which are also used for weather

FIGURE 7.19   MERIS image of May 4, 2004 shows a large sand/dust storm from the Sahara Desert making its way into southern Italy. (*Image courtesy of European Space Agency.*)

monitoring. These near-polar orbiting satellites carry very sensitive light sensors known as the Operational Linescan System (OLS), which can detect light emission from the Earth's surface at night. The OLS sensor provides twice-daily coverage with a swath width of 3000 km. It has two broad bands, that is, a visible and near-IR band (0.4 to 1.1 μm) and a thermal-IR band (10.0 to 13.4 μm). Visible pixels are relative values ranging from 0 to 63 (6 bit) rather than absolute values in Watts/mi^2. IR pixel values correspond to a temperature range of 190° to 310° Kelvin in 256 equally spaced steps

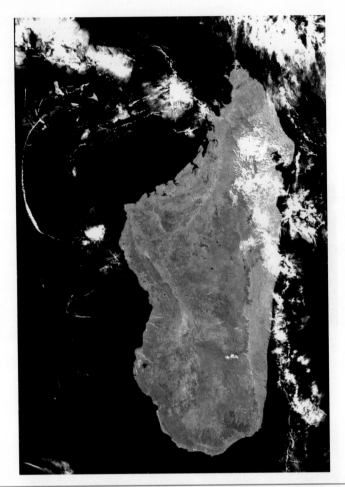

FIGURE 7.20   The MERIS false color image of the East African island of
Madagascar on August 13, 2003. (*Image courtesy of European Space
Agency.*)

(8-bit data). The OLS data can be acquired in either full or smoothed
spatial resolution modes. The full resolution fine data have a spatial resolution of 0.56 km, while onboard averaging of 5×5 pixels
of fine data leads to the smoothed data with a nominal spatial resolution of 2.7 km. The digital data from DMSP-OLS have been
archived at NOAA's National Geophysical Data Center (NGDC)
since 1992, most of which is in the smoothed spatial resolution
mode. The DMSP-OLS has a unique capability to observe faint
sources of visible near-IR emissions present at night on the Earth's
surface, including cities, towns, villages, gas flares, heavily lit

FIGURE 7.21   The nighttime lights of the United States depicts the lights from cities and gas flares (off the coasts of Southern California and Louisiana) in white. Source data collected October 1994–March 1995. (*Image courtesy of NASA.*)

fishing boats, and fires. By analyzing a time series of DMSP-OLS images, it is possible to define a reference set of "stable" lights, which are present in the same location on a consistent basis. Fires, or urban centers, can be detected as lights on the land surface outside the reference set of stable lights. Figure 7.21 shows the lights from U.S. cities and gas flares (off the coasts of Southern California and Louisiana) in white using the DMSP-OLS data collected between October 1994 and March 1995.

In addition to the GOES series of the United States, other countries have also launched geostationary meteorological satellites into orbit to carry out the tasks of monitoring weather and climate. All of these satellite programs join the World Weather Watch project of the World Meteorological Organization. These include geostationary meteorological satellite (GMS) satellite series of Japan, FengYun series of China, INSAT series of India, and the Meteosat satellites by a consortium of European nations.

The Japanese GMS series, also known as its nickname, "Himawari" (meaning a "sunflower"), is on the geostationary orbit at 140°E longitude to cover Japan and nearby regions. Similar to GOES, GMS provides imaging of the Earth every 30 minutes. GMS has two spectral bands: visible IR band in the range from 0.5 to 0.75 μm (1.25-km resolution), and thermal-IR band from 10.5 to 12.5 μm (5-km resolution). After the Himawari-6, the GMS series was replaced by a multifunctional transport satellite (MTSAT) series to broaden its scope of operation. The MTSATs are a series of weather and aviation control satellites providing coverage for the hemisphere centered on 140°E, including Japan and Australia as the principal users of the satellite imagery. They can provide imagery in five bands: one visible band at 1-km resolution and four IR bands at 4-km resolution, including the water vapor channel.

China has launched both polar orbit and geostationary orbit meteorological satellites since 1988. The satellites in the Fengyun-1 series (literally, "wind cloud," abbreviated "FY") are in polar-orbiting, sun-synchronous orbits. The satellites in the FY-2 series are in geostationary orbit. The newer FY-3 series is an improved generation of polar-orbiting, sun-synchronous weather satellites. The FY-3A mission carried three imagers: visible and IR radiometer (VIRR), moderate resolution visible and IR imager (MODI), and microwave radiation imager (MWRI). VIRR is a 10-channel radiometer based on the multi-channel visible and IR scanning radiometer (MVISR, 10 channels) onboard FY-1C/D satellites. MODI has 20 channels mainly located in the visible and near-IR regions, which provide complementary measurements to the VIRR's IR channels. MODI has five channels (four VIS and one thermal-IR) with 250-m spatial resolution, which aims at imaging the Earth with relatively high resolution and near true color imagery during the day and thermal-IR imagery at night. Table 7.11 briefly describes the spectral bands of MODI. The MWRI is a conical scanning microwave imager at six frequency points with dual polarizations (12 channels). This sensor measures thermal and microwave emissions from land and ocean surfaces, water vapors in the atmosphere, and clouds. Since some microwave channels are longer than 1 mm, the imager can penetrate clouds and provide all-weather imaging capability. The spatial resolution of individual MWRI bands varies from 15 to 80 km. The image data from Chinese meteorological satellites are used for monitoring weather systems, disasters, and air quality, in addition to various environmental applications.

The Indian National Satellite System (INSAT) provides a series of multipurpose geostationary satellites launched by the ISRO to satisfy the telecommunications, broadcasting, meteorology, and search and rescue needs of India. INSAT was commissioned with the launch of INSAT-1B in August 1983. Some of the satellites carry a very-high-resolution radiometer (VHRR) and CCD cameras for metrological imaging. KALPANA-1 is an exclusive meteorological satellite launched on September 12, 2002 into a geostationary orbit. It carries VHRR and data relay transponder (DRT) payloads to provide meteorological services. It is located at 74°E longitude. The satellite was originally known as MetSat-1, and was later renamed to Kalpana-1 in memory of Kalpana Chawla, a NASA astronaut who died in the Space Shuttle Columbia disaster. The three band images are visible, thermal-IR, and water vapor IR.

Channel	Central Wavelength (μm)	Band Width (μm)	Spatial Resolution (m)
1	0.470	0.05	1000
2	0.550	0.05	250
3	0.650	0.05	250
4	0.865	0.05	250
5	1.640	0.05	250
6	2.130	0.05	1000
7	0.412	0.02	1000
8	0.443	0.02	1000
9	0.490	0.02	1000
10	0.520	0.02	1000
11	0.565	0.02	1000
12	0.650	0.02	1000
13	0.685	0.02	1000
14	0.765	0.02	1000
15	0.865	0.02	1000
16	0.905	0.02	1000
17	0.940	0.02	1000
18	0.980	0.02	1000
19	1.030	0.02	1000
20	11.50	2.0	250

TABLE 7.11   Spectral Bands of FengYun-3A

The ESA has long been building Europe's orbital weather satellites: the Meteosat series of geostationary spacecraft, the first of which was launched in 1977. The success of the early Meteosats led to the creation of the European Organisation for the Exploitation of Meteorological Satellites (EUMETSAT) in 1986. The first Meteosat second-generation (MSG) satellite was launched on August 28, 2002, and became operational on January 29, 2004 when it was re-designated Meteosat-8. Since then, it has continuously returned highly detailed imagery of Europe, the North Atlantic, and Africa every 15 minutes, for operational use by meteorologists. The second MSG was launched on December 21, 2005. It is currently in the same fixed section of orbital space as MSG-1, that is, in geostationary orbit close to where the equator meets the Greenwich meridian. MSG-2 has 12 spectral bands and improves the spatial resolution for the visible light spectral band to 1 km as opposed to 2.5 km on MSG-1.

7.5   Commercial Satellites

7.5.1   IKONOS

IKONOS, launched on September 24, 1999, is the first commercial Earth observation satellite. IKONOS sensors use pushbroom scanning technology. The detectors at the focal plane include a panchromatic (PAN) and a multispectral (MS) sensor, with 13,500 pixels and 3375 pixels, respectively (cross-track). The imaging sensors collect both 4-m resolution MS and 1-m resolution PAN imagery. The multispectral bands take images in the blue, green, red, and near-IR wavelength regions (Table 7.12). The multispectral images can also be merged with panchromatic images of the same locations to produce "pan-sharpened color" images of 1-m resolution. The IKONOS satellite runs on a polar, circular, sun-synchronous orbit at a height of 681 km. The satellite's altitude is measured by two star trackers and a sun sensor along with an onboard GPS receiver, ensuring that it acquires imagery with very high positional accuracy. The revisit rate for IKONOS is 3 to 5 days off-nadir and 144 days for true-nadir. Both MS and PAN sensors have a swath width of 11 km on the ground. The sensors collect data with an 11-bit (0–2047) sensitivity and are delivered in an unsigned 16-bit (0–65535) data format. From time to time, the data are rescaled down to 8-bit (0–255) to decrease file size. When this occurs, much of the sensitivity of the data needed by remote sensing scientists is lost. IKONOS imagery began being sold on January 1, 2000. Because of extremely detailed spatial information and highly accurate location information, IKONOS images have gained great attention not only in commercial uses, but also in military, mapping, urban, and precision agriculture, among others. Figure 7.22 shows an IKONOS false color composite (bands 2, 3, and 4) of downtown Indianapolis, Indiana, U.S.A. The image was acquired on October 6, 2003. Red color represents vegetation. Buildings and roads are in bluish gray, except for a few buildings that are in white (indicating high reflectance). The river is black. We can see clearly shadows cast by tall buildings in the central business district.

Band	Wavelength Range	Spatial Resolution (m)
1 (Blue)	0.445–0.516 µm	4
2 (Green)	0.506–0.595 µm	4
3 (Red)	0.632–0.698 µm	4
4 (Near-IR)	0.757–0.853 µm	4
PAN	0.45–0.90 µm	1

TABLE 7.12   IKONOS Spectral Bands

FIGURE 7.22   IKONOS false color composite (bands 2, 3 and 4) of the downtown Indianapolis acquired on October 6, 2003.

7.5.2   QuickBird

QuickBird is a high-resolution commercial Earth observation satellite owned by Digital Globe. It was launched on October 18, 2001 as the first satellite in a constellation. By lowering its orbital height to 450 km, QuickBird is capable of collecting images in higher spatial resolutions than IKONOS. QuickBird sensors simultaneously collect panchromatic images at 0.61 m (approximately 2 feet) resolution and multispectral images at 2.5 m (approximately 8 feet) resolution. The multispectral images consist of four bands in the blue, green, red, and near-IR wavelength regions. Table 7.13 details each of the spectral bands. Similar to IKONOS images, the multispectral images of Quick-Bird can be merged with panchromatic images of the same locations to produce "pan-sharpened" color images to 0.61-m resolution. Quick-Bird is in a polar, sun-synchronous orbit (inclination: 98°), with a revisit frequency of 1 to 3.5 days depending on latitude, at 60-cm resolution. The image swath width is 16.5 km at nadir. However, QuickBird has an accessible ground swath of 544 km centered on the satellite ground

Spectral Band	Wavelength (μm)	Resolution (at nadir)	Resolution (at 30° off nadir)
1 (Blue)	0.45–0.52	2.4 m	2.9 m
2 (Green)	0.52–0.60	2.4 m	2.9 m
3 (Red)	0.63–0.69	2.4 m	2.9 m
4 (NIR)	0.76–0.90	2.4 m	2.9 m
Panchromatic	0.45–0.90	0.61 m	0.73 m

TABLE 7.13   Spectral Bands of QuickBird

track due to 30° off nadir viewing. QuickBird has a large onboard storage capacity (128 GB capacity, or approximately 57 single-area images). The satellite's star trackers and onboard GPS allow collecting highly accurate locational data, which is essential for large-scale mapping without using ground control points (GCPs). Quickbird imagery has spatial resolutions comparable to aerial photography. At a resolution of 61 cm, details such as buildings and other infrastructure, cars, and even large individual trees can be easily identified. Figure 7.23 shows a QuickBird image of September 11 Ground Zero, New York, on August 27, 2010. Very-high-resolution imagery, such as QuickBird, can be imported into GIS for extraction of much useful information, and can also be used as a backdrop for mapping applications, such as Google Earth and Google Maps. The 2004 Indian Ocean earthquake occurred

FIGURE 7.23   QucikBird image of September 11 Ground Zero, New York City on August 27, 2010. (*Image courtesy of DigitalGlobe.*)

in the early morning of December 26, 2004, with an epicenter off the west coast of Sumatra, Indonesia. The earthquake was caused by subduction and triggered a series of devastating tsunamis along the coasts of most landmasses bordering the Indian Ocean, killing over 230,000 people in 14 countries, and inundating coastal communities with waves up to 30 m (100 feet) high. It was one of the deadliest natural disasters in recorded human history. Indonesia was hardest hit, followed by Sri Lanka, India, and Thailand. A comparison between the before-event (June 23, 2004) and the after-event (December 28, 2004) QuickBird imagery clearly shows the affected areas (Fig. 7.24).

(a) June 23, 2004

(b) December 28, 2004

FIGURE 7.24   QuickBird images of the 2004 Indian Ocean Tsunami. A comparison between the before (June 23, 2004) and after (December 28, 2004) imagery clearly shows the affected areas. (*Image courtesy of DigitalGlobe.*)

7.6   Summary

Earth observing satellites may be categorized into two broad groups: Earth resources satellites and environmental satellites. The Earth resources satellites are designed for monitoring, assessing, and mapping natural resources, ecosystems, and environments. The images acquired by the Earth resources satellites are characterized by moderate spatial resolution (typically 10 to 100 m) with swath widths of less than 200 km and revisit time of two weeks to one month. Environmental satellites include meteorological and oceanographic satellites, which acquire images with much coarser resolution, typically several hundred to a thousand meters, and much wider swath of hundreds to thousands of kilometers. They often have a very high temporal resolution, covering the entire Earth daily or hourly. For the past several decades, both Earth resources and environmental satellites have contributed to our knowledge of the planet Earth, and have advanced remote sensing science and technology. Images from Earth resources satellites are more commonly used in studies at local and regional scales, while those from environmental satellites are used in regional and global studies. The latter becomes increasingly important in addressing global environmental change concerns. Since 1999, a new category of satellites has become available for public use, that is, commercial satellites. These satellites provide images of ultra-high spatial resolution, often a few meters or even sub-meters, and have stereo imaging capacity. Because commercial satellites are also capable of high locational accuracy, their images are widely used in military, mapping, and urban applications. As these images are becoming routinely available through the Internet and via telecommunication means, new issues of personal privacy and spatial intelligence emerge.

Spaceborne remote sensing is changing rapidly nowadays. The United States, as well as a few other nations in the world, has developed "due use" satellite systems that serve government agencies, military services, and private corporations (Campbell, 2007; Lillesand, Kiefer, and Chipman, 2008). Constellations of satellites, which are often incorporated with a sensor web of satellite systems, *in situ* sensors, and ground control systems, are attracting many interests. This chapter focused on providing readers with key knowledge of Earth observing satellites, as it is essential to acquire knowledge of more sophisticated satellite systems. The satellite systems introduced in this chapter typically operate within the optical spectrum (ranging from 0.3 to 14 μm), which is further divided into UV, visible, near-IR, mid-IR, and thermal-IR. The next chapter will discuss how to process and analyze satellite images in the visible, near-IR, and mid-IR wavelengths. Thermal-IR remote sensing detects emitted radiation instead of reflected energy. A separate treatment of thermal remote sensing is given in

Chap. 9. Many satellites also carry microwave sensors (such as radars). Microwave sensors, along with Lidar sensing, are discussed in detail in Chap. 10.

Key Concepts and Terms

Advanced very-high-resolution radiometer (AVHRR)   The primary sensor onboard the NOAA POES satellites, used for meteorology, and global and regional monitoring. The AVHRR sensor detects radiation in the visible, near-, mid-, and thermal-IR regions with spatial resolution of 1.1 km. AVHRR data can be acquired in four operational modes, differing in spatial resolution and method of transmission.

DMSP-OLS   Defense Meteorological Satellite Program Operational Linescan System is a very sensitive light sensor that can detect light emission from the Earth's surface at night. The OLS sensor provides twice-daily coverage with a swath width of 3000 km. It has two broad bands, that is, a visible and near-IR band (0.4 to 1.1 μm) and a thermal-IR band (10.0 to 13.4 μm).

Earth resources satellites   Satellites designed for monitoring, assessing, and mapping natural resources, ecosystems, and environments. The images acquired are characterized by moderate spatial resolution (typically 10 to 100 m) with swath widths of less than 200 km and revisit time of two weeks to one month.

Enhanced TM plus (ETM+)   The remote sensing system onboard Landsat-7 that acquired visible, near-, mid-, and thermal-IR spectral bands in seven channels and a panchromatic band of 15-m spatial resolution.

Environmental satellites   Meteorological and oceanographic satellites that acquire images with much coarser resolution, typically several hundred to a thousand meters, and much wider swaths of hundreds to thousands of kilometers. They often have a very high temporal resolution, covering the entire Earth daily or hourly.

Geostationary orbit   The altitude of a satellite is high enough so that its orbital period is equal to the Earth's rotation. The satellite is in the equatorial plane.

GOES (geostationary operational environmental satellite)   A key element in NOAA's National Weather Service operations, GOES weather imagery and sounding data provide a continuous and reliable stream of environmental information used to support weather forecasting, severe storm tracking, and other meteorological tasks. The GOES I-M Imager is a five-channel (one visible, four IR) imaging radiometer designed to sense radiant and solar reflected energy from sampled areas of the Earth.

IKONOS   The first commercial Earth observation satellite launched in 1999. The IKONOS sensors collect both 4-m resolution multispectral and 1-m resolution panchromatic imagery in the blue, green, red, and near-IR wavelength regions.

Instantaneous field of view (IFOV)   A measure of the ground area viewed by a single detector element in a given instant in time.

IRS satellites   India successfully operates several Earth resources satellites that gather data in the range from visible, through near-IR to mid-IR bands, beginning in 1988. The third and fourth satellites in the series, IRS-1C and IRS-1D, both carried three sensors: a single-channel panchromatic high-resolution camera, a medium-resolution four-channel LISS, and a coarse-resolution two-channel WiFS.

Landsat   An ongoing series of Earth resources satellites that launched in 1972, and are designed to provide broad-scale repetitive surveys of the landscape. Sensors include RBV, MSS, TM, and ETM+.

Landsat data continuity mission (LDCM)   A collaborated effort between NASA and the USGS that provides continuity with Landsat land imaging data sets, with an earliest projected date of launch in 2012. Its science instruments include the OLI and the TIRS.

MERIS (medium resolution imaging spectrometer)   Launched on January 3, 2002 by the European Space Agency (ESA), MERIS is onboard its polar-orbiting Envisat-1 Earth Observation Satellite. MERIS is primarily dedicated to ocean color observations, but is used in atmospheric and land surface related studies as well. MERIS is a programmable, medium-spectral resolution, imaging spectrometer. The spatial resolution of the sensor is 300 m (at nadir), but this resolution can be resampled to 1200 m onboard.

Moderate resolution imaging spectroradiometer (MODIS)   A key instrument designed to be part of NASA's EOS to provide long-term observation of the Earth's land, ocean, and atmospheric properties and dynamics. Among its 36 spectral bands, the first two bands in the red and near-IR regions have a spatial resolution of 250 m, five additional bands have a spatial resolution of 500 m in the visible to SWIR region, and the remaining 29 spectral bands have a spatial resolution of 1000 m in the mid- and long-wave thermal-IR regions.

Multispectral scanner (MSS)   The remote sensing system onboard Landsat-1, -2, -3, -4, and -5 that acquired visible and near-IR spectral bands in four channels.

Orbit   The path followed by a satellite is referred to as its orbit.

POES satellites   A series of polar-orbiting environmental satellites operated by the U.S. National Oceanographic and Atmospheric Administration (NOAA), starting in 1978. NOAA POES satellites are engineered in sun-synchronous, near-polar orbits, aimed at providing complementary information to the geostationary meteorological satellites. At any time, NOAA ensures at least two active POES satellites operational in orbit, so that it can provide at least four image acquisitions per day for any location on Earth.

QuickBird   A high-resolution commercial Earth observation satellite owned by DigitalGlobe and launched in 2001. Quickbird sensors simultaneously collect panchromatic images at 0.61 m (approximately 2 feet) resolution and multispectral images at 2.5 m (approximately 8 feet) resolution in the blue, green, red, and near-IR wavelength regions.

Return beam vidicon (RBV)   The remote sensing system onboard Landsat-1, -2, and -3 that acquired visible and near-IR spectral bands in three channels.

SPOT (satellite pour l'observation de la terre)   A French satellite program that provides optical imaging of the Earth resources. SPOT satellites use along-track scanning technology. Each satellite carries two high-resolution visible (HRV) imaging systems, capable of sensing either in single-channel panchromatic (PLA) mode, or in three-channel multispectral (MLA) mode.

Sun-synchronous orbit   The orbital-plane satellite moves about the Earth at the same angular rate that the Earth moves around the sun, so that the satellite crosses the equator at the same local sun time on the sunlit side of the Earth.

Swath   As a satellite revolves around the Earth, the sensor "sees" a certain portion of the Earth's surface. The area imaged on the surface is referred to as the swath.

Thematic mapper (TM)   The remote sensing system onboard Landsat-4 and after that acquired visible, near-, mid-, and thermal-IR spectral bands in seven channels.

Review Questions

1.   What are some advantages of satellite images over aerial photographs? Are there any disadvantages?

2.   Explain why the majority of current Earth resources satellites are devised to follow a near-polar orbit instead of a geostationary orbit. Are there any benefits for devising a geostationary Earth resources satellite?

3.   The Landsat program started nearly 40 years ago in 1972. What are some important factors contributing to the success of this long-lasting Earth resources observing program?

4.   The pixel size of Landsat TM imagery is 30 × 30 m, and the dimension of a complete scene is 185 km × 185 km. Calculate the number of pixels for a single band of a TM scene.

5.   SPOT satellite images are widely employed in various application areas outside and inside the United States, in spite of the existence of lower-cost Landsat imagery. Can you explain why SPOT images are equally welcomed in the United States?

6. Comparing with Landsat and SPOT, the MODIS sensor provides only coarse spatial-resolution imagery. However, it has been recently widely accepted by the scientific community. Can you speculate the reasons behind this success?

7. Identify some application areas for NOAA's AVHRR weather satellie, in addition to meteorological studies.

8. Why is NOAA's GOES so useful in tracking severe storms and hurricanes? Can you identify some other natural disasters where meteorological satellites might be valuable?

9. Provide an example of a practical application (e.g., urbanization) in which DMSP-OLS data and Landsat or other Earth resources satellite may be used together.

10. How would commercial satellites such as IKONOS and Quickbird affect our life, as well as decision making that needs geospatial data?

11. Explain how advances in sensor technology have affected our ability to view the Earth from space today and tomorrow.

References

Canada Centre for Remote Sensing. 2007. Tutorial: Fundamentals of Remote Sensing, http://www.ccrs.nrcan.gc.ca/resource/tutor/fundam/chapter2/08_e.php. Accessed on December 3, 2010.

Campbell, J. B. 2007. *Introduction to Remote Sensing*, 4th ed. New York: Guilford Press.

Justice, C. O. et al. 1998. The Moderate Resolution Imaging Spectroradiometer (MODIS): Land remote sensing for global change research. *IEEE Transactions on Geoscience and Remote Sensing*, 36:1228–1249.

Lillesand, T. M., R. W. Kiefer, and J. W. Chipman. 2008. *Remote Sensing and Image Interpretation*, 6th ed. Hoboken, N.J.: John Wiley & Sons.

Smith, L. C., Y. Sheng, G. M. MacDonald, and L. D. Hinzman. 2005. Disappearing arctic lakes. *Science*, 308(5727):1429.

CHAPTER 8

Digital Image Analysis

8.1 Introduction

Digital image analysis aims at detecting, identifying, measuring, and analyzing features, objects, phenomena, and processes from digital remote sensing images, and uses computers to process digital images. In the context of digital analysis of remotely sensed data, the basic elements of image interpretation, although initially developed based on aerial photographs, should also be applicable to digital images. That is to say, manual interpretation and analysis is still an important method in digital image processing. However, the majority of digital image analysis methods are based on tone or color, which is represented as a digital number (brightness value) in each pixel of the digital image. The development of digital image analysis and interpretation is closely associated with availability of digital remote sensing images, advances in computer software and hardware, and the evolution of satellite systems. It has now become a broad subject in remote sensing, and therefore it is impossible to discuss all, or even the majority of the methods and techniques in this chapter. Instead, the basic principles of digital image processing are introduced, without getting into advanced algorithms that often require strong mathematical and statistical backgrounds. Before main image analyses take place, preprocessing of digital images are often required. Image preprocessing may include detection and restoration of bad lines, geometric rectification or image registration, radiometric calibration and atmospheric correction, and topographic correction. To make interpretation and analysis easier, it may also be necessary to implement various image enhancements, such as contrast, spatial and spectral enhancements. Finally, we discuss thematic information extraction from remotely sensed image data. The focus is placed on image classification, largely applicable to multispectral images, and on the unique methods and techniques for hyperspectral images.

8.2   Image Corrections

Geometric and atmospheric corrections are important steps in image preprocessing. Remote sensing images collected from airborne or spaceborne sensors may contain systematic and nonsystematic errors in terms of image geometry. Systematic errors may be caused by scan skew (the ground swath is not normal to the ground track but is slightly skewed, producing across-scan geometric distortion); mirror-scan velocity variance (the mirror scanning rate is not constant across a given scan, producing along-scan geometric distortion); platform velocity (changes in the speed of the platform produces along-track scale distortion); Earth rotation (Earth's rotation results in a shift of the ground swath being scanned, causing along-scan distortion); and the curvature of the Earth's surface. Nonsystematic errors are created when the sensor platform departs from its normal altitude or the terrain increases in elevation, resulting in changes in image scale. In addition, the axis of a sensor system needs to maintain normal to the Earth's surface and the other parallel to the platform's flight direction. If the sensor departs from this attitude, irregular, random geometric distortion may occur. Geometric correction corrects systemic and nonsystematic errors contained in the remote sensing system and during image acquisition (Lo and Yeung, 2002). It commonly involves: (1) digital rectification, a process by which the geometry of an image is made planimetric; and (2) resampling, a process of extrapolating data values into a new grid. During geometric rectification, we need to identify the image coordinates (i.e., row and column) of several clearly discernible points, called ground control points (GCPs), in the distorted image (A1–A4 in Fig. 8.1) and match them to their true

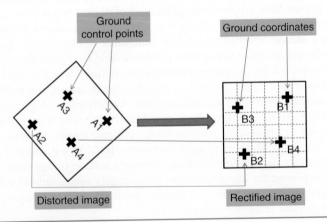

Figure 8.1   Image registration.

positions in ground coordinates (B1–B4 in Fig. 8.1), which may be measured from a map or a previously rectified image. Once GCP pairs are identified, proper mathematical equations are used to transform the coordinates of the distorted image into new coordinates in alignment with their true ground positions. The procedure of resampling is then followed to determine the new pixel values in the new pixel locations of the corrected image based on the original digital pixel values in the uncorrected image. Three methods are frequently used for resampling: (1) nearest neighbor, (2) bilinear, and (3) cubic convolution. The nearest neighbor resampling uses the input cell value closest to the output cell as the assigned value to the output cell (Fig. 8.2). This method is straightforward and easy to compute. Since it does not alter the original data, it may be the most appropriate method if discrimination between land cover or vegetation types or locating the boundaries among the thematic classes is the top priority in a specific application. However, the nearest neighbor method may result in some pixel values being duplicated while others lost.

The bilinear interpolation method estimates the output cell value by taking a weighted average of four pixels in the original image nearest to the new cell location (Fig. 8.3). Comparing with the nearest neighbor method, the bilinear interpolation method produces a smoother image, although it is computationally more expensive. Since this method modifies the original data values, it can affect subsequent image analysis. It is recommended to perform bilinear resampling after image classification (Lillesand, Kiefer, and Chipman, 2008). The cubic convolution method estimates the output cell value by calculating the weighted average of the closest sixteen input cells to the new cell location (Fig. 8.4). This method produces the smoothest image, but it requires the most computational resource and time.

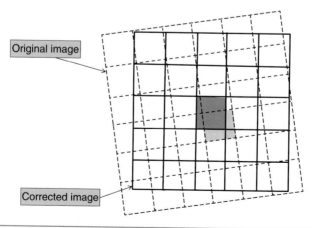

FIGURE 8.2   The nearest neighbor method of image resampling.

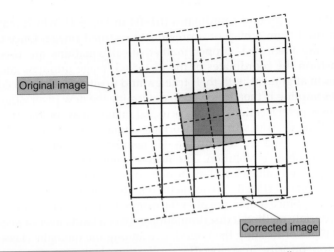

<small>Figure 8.3</small>   The bilinear interpolation method of image resampling.

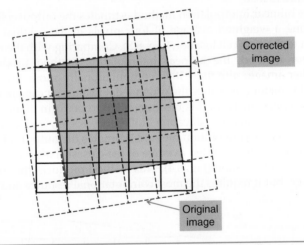

<small>Figure 8.4</small>   The cubic convolution method of image resampling.

Errors or inconsistencies in image brightness values may occur in raw images due to variations in scene illumination, viewing geometry, atmospheric conditions, and sensor noise and response characteristics. These errors or inconsistencies, collectively known as "radiometric noise," vary with the type of sensor and platform used to acquire the data and the conditions during data acquisition. Radiometric corrections aim at reducing the influence of the noise, and facilitating image interpretation and analysis as well as quantitative measurements from remotely sensed images. This section focuses on discussion of radiometric correction for electro-optical sensors because they are more frequently used.

For each data acquisition, the Earth-sun-sensor geometry relationship is different. As a result, variations in scene illumination and viewing geometry between images can be corrected by modeling the geometric relationship and distance between the area of the Earth's surface imaged, the sun, and the sensor (Canada Centre for Remote Sensing, 2007). Image brightness magnitudes must be balanced or "normalized" across and between image scenes and spectral bands, and between different image acquisition dates. This procedure is especially needed to create an image mosaic of a large area that requires multiple image scenes to cover or to study changes in the surface reflectance over time. For such applications, corrections for sun elevation and the Earth–sun distance become important. The sun elevation correction can be implemented by dividing each pixel value in an image by the sine of the solar elevation angle for the time and location of image acquisition. The correction normalizes for the seasonal position of the sun relative to the Earth, assuming the sun was at the zenith when remote sensing occurred (Lillesand et al. 2008). Further, the correction needs to be carried out to account for the seasonal change of the Earth–sun distance, because the irradiance from the sun reduces on the order of the square of their distance (Lillesand et al., 2008).

In Chap. 2, we discuss how the atmosphere affects electromagnetic energy through processes of absorption, scattering, and reflection. Either absorption or scattering reduces the energy illuminating the ground surface. This atmospheric attenuation can be modeled mathematically to compute the amount of energy reaching a ground target, given that the information about atmospheric conditions during the image acquisition is available. Atmospheric scattering introduces "haze" to an image, and thus reduces image contrast. Various methods have been developed for haze compensation. One method is to examine the observed brightness values in an area of shadow or for a very dark object (such as a large, clear lake). The reflectance from these features in the near-IR region of the spectrum is assumed to be zero. If we observe minimum digital number (DN) values much greater than zero in another band over such an area or object, they are considered to have resulted from atmospheric scattering. This offset value can be subtracted from all pixels in that band (Sabins, 1997; Fig. 8.5).

If a single-date image is used for image classification, atmospheric correction may not be required (Song et al., 2001) because signals from the terrestrial features are much stronger than those from the atmosphere. However, if a project requires estimating biophysical parameters from the Earth's surface, such as biomass, sediment, and land surface temperature, subtle differences in the surface reflectance are significant. In that case, conversion of DNs to absolute radiance values becomes a necessary procedure. Each sensor has its own response characteristics and the way in which the analog signal is converted to a digital number (analog-to-digital) conversion. Conversion of DNs to known (absolute) radiation or reflectance units

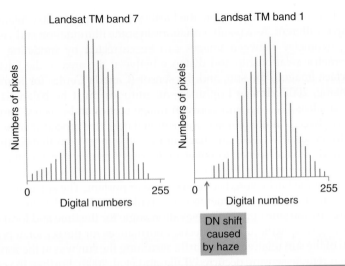

FIGURE 8.5   Histograms for Landsat TM bands 7 and 1. The lack of pixels with low DNs in band 1 is caused by illumination from light selectively scattered by the atmosphere. (*After Sabins, 1997.*)

also facilitates comparison between image data, especially for comparative analysis of several images taken by different sensors (for example, TM sensor on Landsat-5 versus ETM+ sensor on Landsat-7).

When multi-temporal or multi-sensor data are used together for an image classification, correction for atmospheric effects is also necessary. In such applications, relative radiometric correction is often applied to normalize the intensities among different bands within an image and to normalize the intensities of bands of one date to a standard scene chosen by the analyst (Jensen, 2005). The correction can be done by single-image adjustment using histogram adjustment (Fig. 8.5), or by multiple-data image normalization using regression. The latter method relates the DNs of the subject scene to those of the reference image band by band to produce a linear regression equation, and the coefficients and intercept of the equation are then used to compute a normalized image (Jensen, 2005). It should be noted that only the DNs of normalization targets in both scenes are used to develop the regression equation. The normalization targets should be chosen so as not to change in reflectance over time (Eckhardt, Verdin, and Lyford, 1990).

Errors or irregularities that occur in the sensor response or data recording and transmission can cause noise in an image such as systematic striping or banding and dropped lines (Canada Centre for Remote Sensing, 2007). Striping is caused by the variations and drift in the response over time among the detectors, which was common in early Landsat MSS data. The "drift" can be different among the six MSS detectors, causing the same brightness to be represented differently by each detector and thus an overall "striped" appearance (Canada Centre for Remote Sensing, 2007). Dropped lines occur when systematic errors result in missing or defective data along a scan line. The relative radiometric correction procedure is frequently applied to remove detector response error, line dropouts, striping, or line-start problems. On May 31, 2003, the ETM+ Scan Line Corrector (SLC) failed permanently, resulting in approximately 22% of the pixels per scene not being scanned. Although it is still capable of acquiring useful image data with the SLC turned off, particularly within the central part of any given scene, the scientific communities have developed a number of methods to fill in the data gaps (USGS, 2004; Roy et al., 2008; Pringle, Schmidt, and Muir, 2009).

8.3   Contrast Enhancement

Various image enhancement methods may be applied to enhance visual interpretability of remotely sensed data as well as to facilitate subsequent thematic information extraction. Image enhancement methods can be grouped roughly into three categories: contrast enhancement, spatial enhancement, and spectral transformation. Contrast enhancement involves changing the original values so that more of the available range of digital values is used for display and the contrast between targets and their backgrounds is increased (Jensen, 2005).

Remote sensors, especially satellite sensors, are typically designed to detect and measure all levels of reflected and emitted energy emanating from the Earth's surface materials, objects, and features. A diverse range of targets (e.g., forest, deserts, snowfields, water, asphalt surface, etc.) can produce large variations in spectral response. No target can utilize all brightness ranges of a sensor. When an image is displayed, it is common that the range of brightness values does not match with the capability of a color display monitor. Thus, a custom adjustment of the range and distribution of brightness values is usually necessary. The range of brightness values present on an image is referred to as contrast. Contrast enhancement is a method of image manipulation that makes the interested image features stand out more clearly by making optimal use of the colors available on the display or output device (Jensen, 2005).

A variety of methods has been developed for contrast enhancement. Three of the methods—linear contrast stretch, histogram equalization, and special stretch—are described next. Before the detailed discussion, we have to know an important concept called "histogram." An image histogram is a graphical representation of the brightness values that comprise an image, with the brightness values displayed along the x-axis of the graph and the frequency of occurrence of each of the brightness values on the y-axis (Fig. 8.6a). By applying different

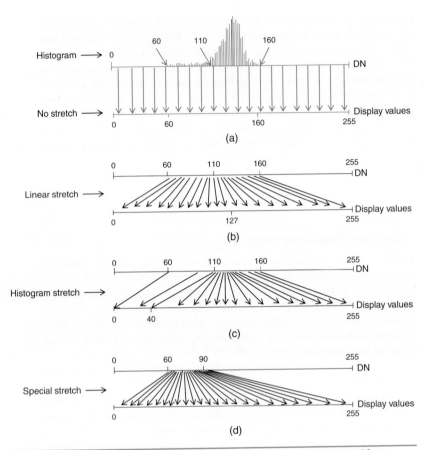

FIGURE 8.6   Commonly used methods of image contrast enhancement. (*After Lillesand et al., 2008.*)

methods of contrast enchantment, the image histogram will change accordingly. Linear contrast stretch linearly expands the original brightness values of the image to the full range of sensitivity of the display device. In our example (Fig. 8.6b), the original minimum brightness value in the histogram is 60, and the maximum brightness value is 160. These 100 levels occupy less than one-half of the full 256 levels available for display. The values of 0 to 59 and 161 to 255 are not displayed. A linear stretch uniformly expands this small range to cover the full range of values from 0 to 255 by matching the old minimum value (60) to the new minimum value (0) and the old maximum value (160) to the new maximum value (255) and by scaling proportionately the old intermediate values between the new minimum and maximum values. This enhancement will make light-toned areas in the image appear lighter and dark-toned areas appear darker, so that visual interpretation is easier and the subtle differences in spectral response are more obvious.

Linear contrast stretch does not work well if we want to see more details in the brightness values from 60 to 160 because there are more pixels in this range. A method called histogram equalization stretch can be more effective. This stretch assigns more display values to the more frequently occurring portions of the histogram and less display values to the lower frequency of brightness values. All pixel values of the image are redistributed so that approximately an equal number of pixels will occupy each of the user-specified output ranges (Jensen, 2005). As seen in Fig. 8.6c, the details in the high frequently occurring brightness values are better enhanced relative to those areas of the original histogram where values occur less frequently. This enhancement method will reduce the contrast in very light or dark parts of the image associated with the tails of a normally distributed histogram (Jensen, 2005).

For a specific application, it may be desirable to assign the entire display range to a particular range of the image values. For example, vegetation features occupy only the digital values from 60 to 90, a very narrow range of the image histogram. We can enhance the characteristics of vegetation features by stretching that narrow range to the full display range from 0 to 255 (Fig. 8.6d). All pixels below or above these values would be assigned either to 0 or 255, and the details in these areas would be hidden. However, the brighter and greater contrast in vegetation features would be useful for delineation and analysis of vegetation.

8.4   Spatial Enhancement

While contrast enhancement focuses on optimal use of the colors available on the display or output device to match with image brightness values collected by a sensor, spatial enhancement is a group of image enhancement techniques that emphasize the spatial relationship of image pixels. These techniques explore the distribution of pixels of varying brightness over an image by defining homogeneous regions and by detecting and sharpening boundary discontinuities between adjacent pixels with notable differences in brightness values. The resultant images are often quite distinctive in appearance, and thus facilitate human recognition.

Various algorithms of spatial enhancement, such as spatial filtering, edge enhancement, and Fourier analysis, use the concept of *spatial frequency* within an image. Spatial frequency is the manner in which brightness values change relative to their neighbors in an image, which relates to the concept of texture that was discussed in Chap. 4. Image areas of high spatial frequency, where the changes in brightness are abrupt over a small number of pixels, are of "rough" texture (Fig. 8.7). In contrast, "smooth" areas with little variation in brightness values over a relative large number of pixels have low spatial frequency. Many natural and man-made features in images have low

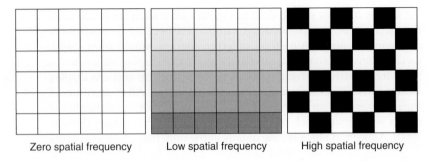

Zero spatial frequency   Low spatial frequency   High spatial frequency

FIGURE 8.7   Examples of image spatial frequencies.

spatial frequency, such as water bodies and large agriculture fields, while others exhibit high spatial frequency, such as geologic faults, edges of lakes, roads, and airports. Spatial filtering is the method that uses user-defined filters to emphasize or suppress image data of different spatial frequencies. Low-pass filters are designed to emphasize larger, homogeneous areas of similar brightness values and suppress local details in an image. In other words, low-pass filters are used to smooth (or to blur) the appearance of an image. High-pass filters, on the other hand, are designed to sharpen the appearance of local details in an image. A high-pass filter may be generated by subtracting a low-pass filter from the original image, leaving behind only the high spatial frequency information.

Spatial filtering involves passing a window of a few pixels in dimension (e.g., 3 pixels × 3 pixels) over each pixel in the image, applying a mathematical operation to the pixel values within that moving window, and creating a new image where the value of the central pixel in the window is the result of the mathematical operation (Fig. 8.8). A simple example of the moving window is the mean

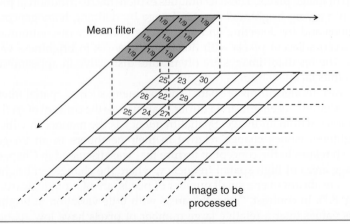

FIGURE 8.8   Spatial filtering using a moving window.

filter, with which a new image will have its pixel value as the average value of the closest nine pixels in the original image. In fact, spatial filtering belongs to a group of generic image processing operations called convolution, which has found itself numerous applications in remote sensing image analysis. The window (also known as kernel), a convolution operator, may be considered as a matrix (or mask) of coefficients that are to be multiplied by image pixel values to derive a new pixel value for a resultant enhanced image. The matrix may be of any size in pixels (e.g., 3×3, 5×5, 7×7) and different shapes (e.g., square, octagonal, nearest neighbor cross; Jensen, 2005). By changing the mathematical operations and the weightings (i.e., coefficients) of the individual pixels in the moving window, a variety of filters can be created to enhance or suppress specific types of features in the image. Cubic convolution, a resampling method as introduced in Sec. 8.2, is an example of the convolution technique. A more complex example involves making the kernel coefficients agree with a statistical distribution, such as a Gaussian distribution (Rajasekar and Weng, 2009).

If we add a high-pass filter to the original image, a high-frequency enhancement will be achieved and an edge-enhanced image will be produced. Linear features (roads, geologic faults, etc.), composed of edges, are often not immediately recognizable from an image. Edge enhancement techniques are designed to highlight linear features, by expanding the width of linear features and by increasing the brightness difference across the features (Sabins, 1997). Digital filters, designed to enhance features oriented toward specific directions, are known as directional filters. These filters are especially useful in geological studies such as detecting linear geologic structures, and are frequently applied to radar images for increasing the visibility of linear features or lineaments (Arnold, 2004). Directional first differencing is one of the directional filtering techniques (Avery and Berlin, 1992; Lillesand et al., 2008). The other category of edge-enhancement filters is non-directional. To this category belongs the Laplacian filter, the most popular non-directional filter. They enhance linear features in all directions, while retaining the original image quality. The only exception is that those linear features oriented parallel with the direction of filter movement are not enhanced (Sabins, 1997).

Any remote sensing image may be considered to consist of several dominant spatial frequencies. We can use a mathematical technique called *Fourier Transform* to separate an image into its various spatial-frequency components. If the pixel brightness values (DNs) in an image are plotted along each row and column, we will see numerous "peaks and valleys," corresponding to the high and low DN values. Therefore, an entire image looks like a complex wave, but mathematically, it can be broken down into simple sine and cosine waves of varying frequency, amplitude, and phase. We may then select and modify its components to emphasize certain groups of frequencies relative to others, and re-combine the image components into an enhanced image at the end. The modification is done through frequency convolution (not the same as the kernel convolution), which is a simple multiplication of an image mask designed by an image analyst, multiplied by a component image. Fourier transform has wide applications in noise removal from an image and image restoration.

8.5   Spectral Transformations

Spectral transformation refers to the manipulation of multiple bands of data to generate more useful information, and involves such methods as band ratioing, vegetation indices, principal component analysis, and so on. Band ratioing is often used to highlight subtle variations in the spectral responses of various surface covers, which may be masked by the pixel brightness variations from individual bands or in standard color composites. Ratioing is accomplished by dividing the DN value of one band by the value of another band. This quotient yields a new set of numbers that may range from 0 to 255, but the majority of the quotients will be fractional (decimal) values between 0 and typically 2 to 3 (Short, 2009). It is important to stretch the resultant ratio image for a better visualization, based on the computer display limits. It is common to multiply the quotients by a normalization factor of 162.3 for an 8-bit display (Lillesand et al., 2008). Brighter tones in the ratio image indicate larger differences in spectral response between the two bands. An effect of band ratioing is that variations in scene illumination because of topographic effects are minimized. The image brightness values are lower in the shadowed side of a hill, but are higher in the sunlit side (Fig. 8.9). However, the ratio values for a specific feature will always be highly similar regardless of the illumination conditions. It is important to remove image noise and atmospheric haze before implementing image ratioing. This is because that the ratioing process cannot remove these additive factors. With a multispectral dataset, we can produce various forms and a number of ratio combinations. For example, with a Landsat MSS dataset of four spectral bands, a total of six ratioed images (band 4/5, 4/6, 4/7, 5/6, 5/7, 6/7) and six reciprocals (band 5/4, 6/4, 7/4, 6/5, 7/5, 7/6) can be generated. In addition to inter-band ratios, an individual band can also be divided by the mean value of all bands in an image dataset, which will result in a normalized band of the particular band (Sabins, 1997). The utilization of any ratio has to consider the particular reflectance characteristics of the features involved and the application of a specific project (Lillesand et al., 2008).

Various forms of ratio combinations have been developed as vegetation indices for assessing and monitoring green vegetation conditions. Healthy green vegetation reflects strongly in the near-IR portion of the electromagnetic spectrum, while absorbing strongly in the visible red spectrum. This strong contrast provides the basis for developing quantitative indices of vegetation conditions.

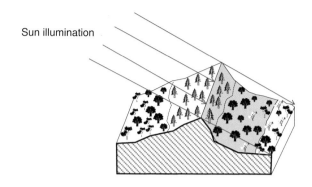

Land Cover/Illumination	Band 1 DN	Band 2 DN	Ratio (Band 1/Band 2)
Deciduous sunlit	20	24	0.83
Deciduous shadow	17	20	0.85
Coniferous sunlit	14	18	0.78
Coniferous shadow	10	13	0.77

Figure 8.9   An illustration of the effect of band ratioing. Variations in scene illumination because of topographic effects are minimized. (*After Lillesand et al., 2008.*)

A vegetation ratio image, for example, Landsat TM band 4 (near-IR infrared—0.76 to 0.90 μm) divided by band 3 (red—0.63 to 0.69 μm), would result in values much greater than 1.0 for healthy vegetation, but a value close to 1.0 for soil and water. Soil and water yield similar brightness values in both the near-IR and red bands. We can also use vegetation indices to discriminate between stressed and non-stressed vegetation. Vegetation under stress will have a low near-IR reflectance, and therefore the ratio values will be smaller than healthy green vegetation. The most popular form of vegetation index is the normalized difference vegetation index (NDVI), which is produced by dividing the difference between the two bands (near-IR and red) by their sum. The NDVI value ranges from −1 to 1, with 0 representing the approximate value of no vegetation and negative values representing non-vegetated surfaces. NDVI has been used in a variety of contexts to assess green biomass and as a proxy to overall health of ecosystems. In Chap. 7, we illustrate two applications of NDVI, one for monitoring global vegetation conditions using a MODIS sensor (Fig. 7.10) and the other for monitoring vegetation conditions in cropland and pasture regions throughout the growing season in the United Sates by using NOAA-18 NDVI weekly composites (Fig. 7.15).

Different bands of a multispectral data set often contain similar information. If you compare individual TM bands of the Terre Haute image (Fig. 3.12 in Chap. 3), you will note that band 1 and band 2 images look similar, as do bands 5 and 7 images. However, band 4 is unique, showing strong contrast with other bands, especially with band 1. These visual and numerical similarities in brightness indicate that some bands are highly correlated. Therefore, there are some redundancies in a multispectral dataset. A statistical technique, known as principal component analysis (PCA), can be used to reduce the data redundancy so that more efficient data analysis and easier interpretation is possible. To explain this mathematical transformation, we can plot pixel values from two bands on a scatter diagram (Fig. 8.10). The transformation defines a new axis (Axis 1) oriented in parallel with the long dimension of the scatter and the second axis perpendicular to Axis 1. These new axes define the directions of principal components, and the principal component image data values are computed as a linear combination of pixel values in the original coordinate system. The same transformation may be applied to a multispectral dataset of any number of bands, or a hyperspectral image. The process typically results in fewer principal components uncorrelated with one another, which are ordered in terms of their explanatory power (higher-order components explain more information; Fig. 8.11). Noise is often denoted as a less-correlated component image (Sabins, 1997). Figure 8.12 illustrates the result of principal component analysis derived from a hyperspectral image. The first component resembles a black-and-white aerial photograph, revealing the information on albedo (the ratio of reflected radiation to incident radiation upon a surface). The second component is mainly associated with healthy vegetation, while the third component is associated with bare soils and grounds. Principal components are usually easier

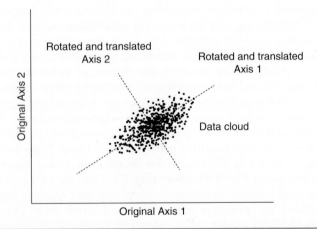

FIGURE 8.10   Rotation of axes in the principal component transformation (example of 2D data).

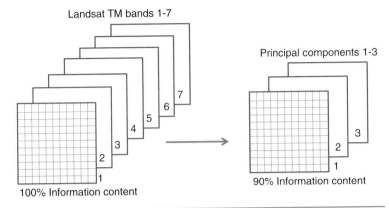

FIGURE 8.11 Illustration of principal component analysis for a Landsat TM image.

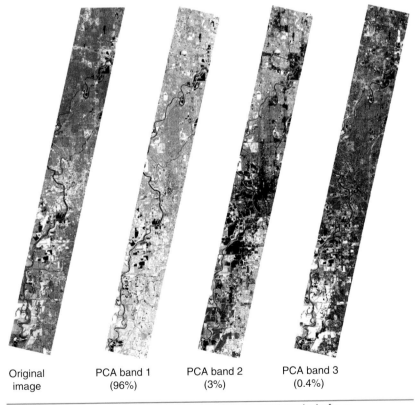

| Original image | PCA band 1 (96%) | PCA band 2 (3%) | PCA band 3 (0.4%) |

FIGURE 8.12 An illustrated result of principal component analysis for a hyperspectral image.

to interpret visually than the original bands. Digital analysis of these component images are also found more efficient and effective. PCA has been used frequently either as a spectral enhancement technique to enhance visual interpretation or as a feature extraction technique for subsequent use in digital classification procedures.

8.6   Image Classification

8.6.1   Image Classification Process

Different features usually have different spectral response characteristics, which produces a unique combination of digital numbers in various bands for each pixel. Image classification uses spectral information represented by the DNs in one or more spectral bands and attempts to classify each individual pixel based on the spectral information. The objective is to assign all pixels in the image to particular classes or themes (e.g., water, vegetation, residential, commercial, etc.) and to generate a thematic "map." Depending on the goal of a specific project, an information class may be divided further into subclasses. For example, vegetation can be broken down into forest, cropland, grassland, pasture, and so on. It is important to differentiate between information classes and spectral classes. The former refers to the categories of interest that the analyst is actually trying to identify from the imagery, while the latter are the groups of pixels that are uniform (or near similar) with respect to their brightness values in the different spectral channels of the data. The ultimate goal of image classification is to derive information classes of interest. Although some digital image processing techniques may be applicable to both multispectral and hyperspectral imagery, we focus our discussion on multispectral imagery in this section and save hyperspectral image analysis for the next section.

Feature extraction is sometimes necessary for subsequent image classification. It serves as a pre-processing step to reduce the spectral, or spatial, dimensionality of an image. Feature extraction can be accomplished by using any type of spectral or spatial transforms to reduce the data or to enhance its multispectral features, or even by simply selecting a subset of multispectral bands. Many potential variables may be used in image classification, including spectral signatures, vegetation indices, transformed images, textural or contextual information, multi-temporal images, multi-sensor images, and ancillary data. Because of different capabilities in spectral separability, the use of too many variables in a classification procedure may decrease classification accuracy (Price, Guo, and Stiles, 2002). It is important to select only the variables that are most effective for separating thematic classes. Selection of a suitable feature extraction approach is especially necessary when hyperspectral data are used. This is because of the huge amount of data volume and high correlations that exist among the bands of hyperspectral imagery, and also because a large number of training samples is required in image classification. Many feature extraction approaches have been developed, including principal component analysis, minimum noise fraction transform, discriminant analysis, decision boundary feature extraction, non-parametric weighted feature extraction, wavelet transform, and spectral mixture analysis (Lu and Weng, 2007).

Before starting an image classification procedure, it is often helpful to generate color composites of various combinations of bands. A spacecraft (Landsat, SPOT, etc.) collects multiple bands of images in black and white, not in color. An interpreter can only be able to distinguish between 10 and 20 shades of grey; however, he or she may be able to distinguish many more colors. Color composites help to identify various features and make interpretation easier. In displaying a color composite image on the screen, computers use the light of three primary colors (red, green, and blue), just like color TVs. The three-color lights can be combined in various proportions in order to produce various colors (Fig. 8.13). A color composite image is generated by assigning each primary color to a separate spectral band (Fig. 8.14). Depending on how well the colors in the image composites match with colors we see in nature, we may produce false and true/natural color composites. A common form of false color composite is generated by projecting a green band image through the blue color gun, a red band through the green color gun, and a near-IR image through the red color gun (Fig. 8.14). A false color composite is valuable to interpret vegetation. Vegetation appears in different shades of red due to its high reflectance in the near-IR band, and the shades vary with the types and conditions of the vegetation (Liew, 2001). In a true

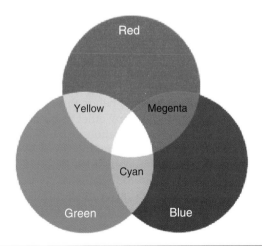

FIGURE 8.13 Three primary colors (red, green, and blue) and other colors by combining different intensities of these three colors.

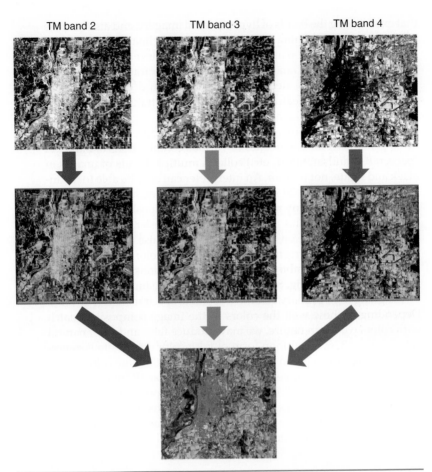

FIGURE 8.14   Process for creating a false color composite using Landsat TM bands.

or natural color composite image, colors resemble a visible color photograph, that is, vegetation in green, water in blue, soil in brown, and so on (Fig. 8.15). A true/natural color image can be generated by projecting blue, green, and red spectral bands through the blue, green, and red color guns, respectively. Many other forms of false and natural color composites via other combinations of bands and color lights may bring out or call attention to individual scene features that are usually present in more subtle expressions (Short, 2009). Furthermore, ratio images can also be used to generate color composites by combining three black-and-white ratio images. This type of color composite favors incorporating more than three bands and facilitates the interpretation of subtle spectral differences (Lillesand et al., 2008). Similarly, it is common to form a color composite using three of the resultant principal component images.

FIGURE 8.15   A comparison of a true/natural color with false color composite.

Generally, there are two approaches to image classification: super-vised and unsupervised classification. In a supervised classification (Fig. 8.16), the analyst identifies in the imagery homogeneous representative samples of different cover types (information classes) of interest to be used as training areas. Therefore, the image analyst

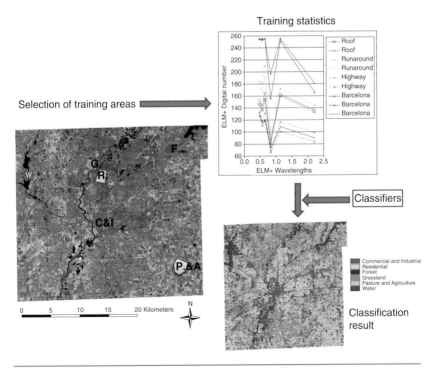

FIGURE 8.16   The process of a supervised image classification.

should be familiar with the geographical area and the actual surface cover types present in the image by experience with thematic maps or by on-site visits. In this sense, the analyst is "supervising" the classification of a set of land cover types. For each thematic class, multiple training areas can be selected. The spectral signature will then be determined for the class based on the DNs for each band from all the pixels enclosed in the train areas. Once numerical spectral signatures have been developed for all the thematic classes, each pixel in the imagery would then compare spectrally with these signatures to determine to which information class it should belong. Various algorithms of supervised classification may be used to assign an unknown pixel to one of a number of the thematic classes, among which are the minimum distance, parallelepiped, and maximum likelihood. In an unsupervised classification, spectral classes are first grouped based solely on digital numbers of the band selected for use (Fig. 8.17). Various statistical methods can be used to cluster all pixels in an image into spectral groups. The analyst needs to specify the number of groups to be looked for in the image. Later, these spectral groups can be combined, disregarded, or broken down into smaller groups. Because the identities of the spectral groups are not initially known,

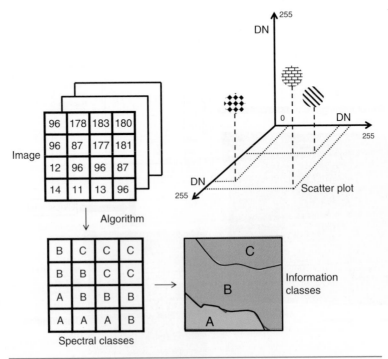

FIGURE 8.17   An illustration of the process of an unsupervised image classification.

the analyst requires matching them with information classes by using some form of references (e.g., a thematic map or large-scale aerial photographs).

8.6.2   Common Image Classifiers

An image classifier is a computer program that implements a classification procedure (Campbell, 2007). Various classifiers of supervised classification have been developed for assigning an unknown pixel to one of the information classes. Among the most frequently used are the minimum distance, parallelepiped, and maximum likelihood classifiers.

The minimum distance classifier computes the mean vectors of all bands for each class based on the training data sets. An unknown pixel is classified by computing the distance between the DN value of this unknown pixel and each mean vector. The unknown pixel is then assigned to a class to which it has the minimum distance in the spectral space. In Fig. 8.18, pixel 1 is assigned to class A. An analyst may specify a threshold value of distance for each class. If a pixel shows a distance from the mean vectors of all classes farther than the specified thresholds, it will be classified as "unknown." The minimum distance classifier is computationally simple, but it is not sensitive to different degrees of variance in the spectral response data. Therefore, the classifier is not applicable where spectral classes are close in the feature space and are of high variance.

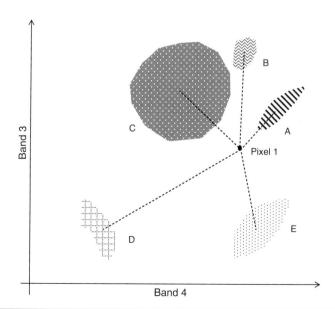

Figure 8.18   An illustration of the minimum-distance-to-means classifier (Pixel 1 is categorized to belong to class A).

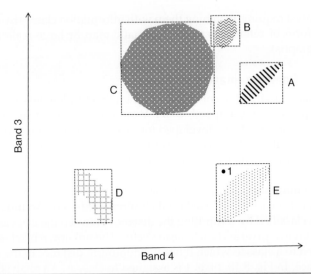

FIGURE 8.19   An illustration of the parallelepiped classifier (Object 1 is categorized as class E). Note: Class B and class C are spectrally overlapped.

The parallelepiped classifier first calculates the range of DN values in each category of training data sets. An unknown pixel is assigned to a class in which it lies. If a pixel is located outside all class ranges, it will be classified as "unknown." In Fig. 8.19, pixel 1 will be assigned to class E. The parallelepiped classifier also has the advantage of being computationally simple. Its main difficulty arises when two or more category ranges overlap. Unknown pixels that lie in the overlapped areas will be assigned arbitrarily to one of the overlapping classes or be labeled as "unknown." The overlapping in the feature space often indicates that two classes correlate spectrally or have high covariance, tendency of spectral values to vary similarly in two bands (Lillesand et al. 2008). It may also be caused by bad selection of training sets for one or more classes.

The maximum likelihood classifier assumes that the training data statistics for each class in each band are normally distributed, that is, the Gaussian distribution. Under this assumption, the distribution of spectral response pattern of a class can be measured by the mean vector and the covariance matrix. A probability density function value is calculated for each class according to these parameters. On a 2D scatter plot, the class DN distribution gives rise to elliptical boundaries, which define the equal probability envelope for each class. Figure 8.20 shows equal probability contours for each class. To classify an unknown pixel, the maximum likelihood classifier computes the probability value of the pixel belonging to each class, and assigns it to the class that has the largest (or maximum) value.

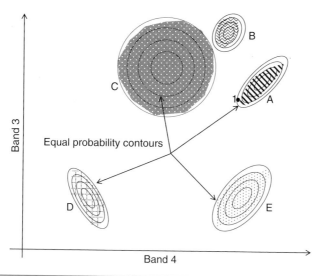

FIGURE 8.20 An illustration of the maximum likelihood classifier (Object 1 falls within the contour of class A, and is categorized as class A).

A pixel will be labeled "unknown" if the probability values are all below the threshold values set by the analyst. This classifier is the most accurate one among the three classifiers we introduce here because it evaluates both the variance and the covariance of the category spectral response patterns. However, if the assumption of data normality is not met for any class, the accuracy of classification result will suffer.

Many advanced classification approaches, such as artificial neural network, fuzzy-set, and expert systems, have been developed for image classification. A description of each category is beyond the scope of this discussion. Readers who wish to have a better knowledge of them may like to look up the article by Lu and Weng (2007). One classification method—object-oriented classification—has recently become quite popular (Gitas, Mitri, and Ventura, 2004; Walter, 2004). Two steps are involved in the classification process (Fig. 8.21). Image segmentation merges pixels into objects, and a classification algorithm is then applied to the objects instead of to the individual pixels. In the process of creating objects, the analyst must control the segmentation scale, which determines the occurrence or absence of an object class, and the sizes of objects. Both parameters affect a classification result. This approach has proven to be able to provide better classification results than per-pixel-based classification approaches, especially for fine spatial resolution data (mostly better than 5 m spatial resolution, such as IKONOS and QuickBird images).

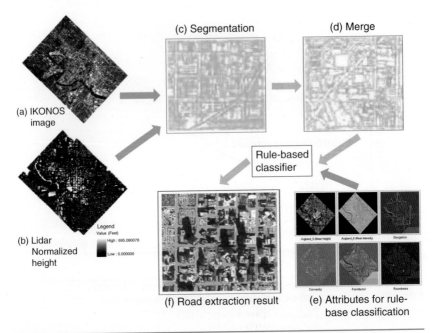

(c) Segmentation

(d) Merge

(a) IKONOS image

Rule-based classifier

(b) Lidar Normalized height

Legend
Value (Feet)
High : 695.080078
Low : 0.000000

(f) Road extraction result

(e) Attributes for rule-base classification

FIGURE 8.21   The process of an object-based image classification for road extraction by using the combined datasets of IKONOS image and Lidar data. The road extraction process includes mainly three steps: segmentation, merging, and rule-based extraction.

8.7   Hyperspectral Image Analysis

In Chap. 3, we discussed the main difference between a multispectral and a hyperspectral imaging system. While multispectral sensors typically provide discrete measurements of average reflected or emitted radiance over isolated, wide spectral bands, hyperspectral sensors produce contiguous, high spectral-resolution radiance spectra. Therefore, multispectral images can only discriminate general differences among surface material types, but a hyperspectral sensor has the potential for detailed identification of materials and better estimates of their abundance. Many techniques for hyperspectral image analysis are derived from the field of spectroscopy, which was developed for mineral mapping in the early 1980s (Goetz et al., 1985). The molecular composition of a particular material results in the distinctive patterns in which the material absorbs and reflects energy at individual spectral wavelengths (Lillesand et al., 2008). Many minerals have unique spectral reflectance signatures, as do soils. Plants display diagnostic "peaks" and "valleys" along their spectral reflectance curves because of the presence of pigments, water, and other chemical constituents. The large number of very narrowly sampled hyperspectral bands allows the use of the remote sensing data to compare with and to replace the expensive measurements obtained in laboratories or from fields. These measurements are often stored in the digital "libraries" of spectral reference data, primarily for minerals, soils, vegetation types, and man-made materials, and are used as reference data in hyperspectral image analysis.

Various techniques have been developed for spectral matching between image spectra and the reference spectra. The spectral angle mapper (SAM; Kruse, Lefkoff, and Dietz, 1993) is a commonly used method for comparing an image spectrum to a reference spectrum. This method treats both spectra as vectors in a multidimensional space, and the number of dimensions equals the number of spectral bands (Fig. 8.22). To compare the two spectra, the spectral angle between them is computed. If the angle is smaller than a set threshold, the image spectrum is considered to match with the reference spectrum. This method does not consider the illumination condition, which modulates the vector length. The result of the SAM is an image showing the best match at each pixel.

Spectral mixture analysis (also known as spectral unmixing) is another technique widely applied in analyzing hyperspectral remote sensing images. Most Earth surfaces are composed of various materials, resulting in mixing of spectral signatures within a single pixel. Depending on how the materials are arranged on the surface, the mixing of the signatures can be represented as a linear model, or a more complex, non-linear model. Spectral mixture analysis assumes that mixed spectral signatures are caused by the mixture of a set of "pure" materials, called end members (Fig. 8.23), whose spectra can be measured in the laboratory, in the field, or from the image itself. Once the spectral mixing model is determined, it can be used to invert the proportions of those "pure materials." The outputs of spectral unmixing are "abundance" images for each end member, which show the fractions of each pixel occupied by each end member. Because the maximum number of end members that can be included is directly

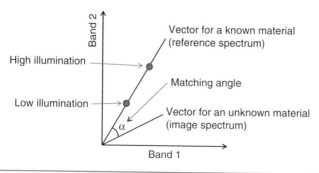

FIGURE 8.22   An illustration of spectral angle mapping concept.

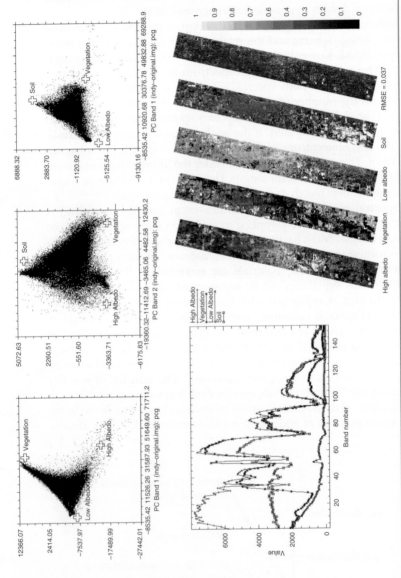

Figure 8.23 Scatterplots of the principal components showing the locations of potential endmembers. Spectral reflectance characteristics of the selected endmembers are also shown. Fraction images (high albedo, low albedo, vegetation, and soil) and error images from spectral mixture analysis of the ALI and Hyperion images.

proportional to the number of spectral bands plus one, the vastly increased dimensionality of a hyperspectral sensor allows deriving a great deal of "pure" materials, and selects only those bands with high SNRs or with less atmospheric absorption effects (Lillesand et al., 2008).

Spectral transform techniques, such as ratioing, vegetation indices, and principal component analysis, have been widely applied in hyperspectral image analysis. Because of the huge number of spectral bands, we have great flexibility to choose spectral bands for use in ratioing. With multispectral data, we have only one choice of using the red and near-IR bands for generating vegetation ratio, while with hyperspectral data, we can have many choices for producing such red and near-IR narrow band combinations (Gong, Pu, Biging, and Larrieu, 2003). Conventional multispectral classification techniques, such as maximum likelihood classifier, cannot be applied to hyperspectral image data due to the high dimensionality of the data. These classifiers require that every training set must include at least one more pixel than there are bands in the sensor, and that it is generally desirable to have the number of pixels per training set be between 10 and 100 times as large as the number of sensor bands to derive statistically meaningful training sets (Benediktsson, Sveinsson, and Arnason, 1995; Lillesand et al., 2008). To overcome these problems, feature extraction becomes necessary before image classification. Feature extraction techniques, such as principal component analysis, or its noise-adjusted version, maximum noise fraction, have been applied in transforming and reducing the data dimension by maximizing the ordered variance of the whole data set or the ratio of between-class variance and within-class variance of the training samples (Pu and Gong, 2011). The maximum noise fraction transform maximizes the SNR in choosing principal components (Green, Berman, Switzer, and Craig, 1988). Principal components resulted from these transforms can be used in subsequent image classification and for determining end member spectra for spectral mixture analysis. Alternatively, a non-parametric classifier, such as a neural network, can be used for classification of hyperspectral images (Filippi and Jensen, 2006).

8.8. Summary

Digital image processing goes far beyond what we have discussed in this chapter. In fact, it is a common interest of multiple fields, such as remote sensing, pattern recognition, signal processing, and computer vision. From a remote-sensing perspective, digital image processing copes with subtle variations in brightness values (DNs). Thus, we are concerned about image data characteristics, especially radiometric, spatial, and spectral qualities. Since most techniques of image corrections and enhancements, as well as feature extraction techniques, do alter image brightness values, we should always select carefully before implementing any technique. For a long time, image classification mainly utilized brightness values of individual pixels in multispectral bands. These types of image classifications have been termed as "spectral classifiers" (Campbell, 2007) or "spectral pattern recognition" (Lillesand et al., 2008). As high spatial-resolution data become readily available today, image texture has been used in image classification, as well as contextual information, which describes the association of neighboring pixel values. In this chapter, we introduced a new classification method known as "object-based image analysis." The stereoscopic viewing capacity of the high-resolution images, such as IKONOS and QuickBird, further stimulates the use of manual interpretation methods and photogrammetric techniques parallel with digital image processing methods (Jensen, 2005; Tong, Liu, and Weng, 2010). Similarly, since the mid-1980s, various hyperspectral sensors continue to emerge, which provide a large amount of hyperspectral image data for applications in different fields. The birth of a unique set of techniques for hyperspectral remote sensing, such as those discussed in this chapter, become necessary. All methods and techniques discussed in this chapter are suitable for both reflected and emitted bands of digital images. However, there are some particular methods suited only for the emitted bands, which will be discussed in Chap. 9.

Key Concepts and Terms

Absolute radiometric correction   Conversion of image DNs to absolute radiance values to account for a specific sensor's response characteristics and the analog-to-digital conversion.

Band ratioing   Dividing the DN value of one band by the value of another band.

Bilinear interpolation resampling   Calculates the output cell value by calculating the weighted average of the four closest input cells based on distance.

Color composite image   Image generated by assigning each primary color to a separate spectral band.

Contrast enhancement   The histogram of DN values is modified so that more of the available range of digital values is used for display, and the contrast between targets and backgrounds is increased.

Cubic convolution resampling   Calculates the output cell value by calculating the weighted average of the closest 16 input cells based on distance.

Digital image analysis   The tasks for detecting, identifying, measuring, and analyzing features, objects, phenomena, and processes from remote sensing images using computers.

Directional filters   Filters designed to enhance features oriented toward specific directions.

False color composite   Commonly generated by projecting a green band image through the blue color gun, a red band through the green color gun, and a near-IR image through the red color gun. A false color composite is valuable to interpret vegetation.

Fourier transform   A mathematical technique that separates an image into its various spatial frequency components. The components are modified through frequency convolution to emphasize certain groups of frequencies and recombine the image components into an enhanced image at the end. Fourier transform has wide applications in noise removal from an image and image restoration.

Geometric correction   Image preprocessing that corrects systemic and nonsystematic errors contained in the remote sensing system and during image acquisition. It commonly involves digital rectification and resampling.

Ground control point (GCP)   A specific pixel on an image or location on a map whose geographic coordinates are known.

High-pass filters   Filters designed to sharpen the appearance of local details in an image.

Histogram equalization stretch   This stretch assigns more display values to the more frequently occurring portions of the histogram and less display values to the lower frequency of brightness values. All pixel values of the image are redistributed so that approximately an equal number of pixels will occupy each of the user-specified output ranges.

Image classification   Uses spectral information represented by the DNs in one or more spectral bands and attempts to assign all pixels in the image to particular classes or themes (e.g., water, vegetation, residential, commercial, etc.) and to generate a thematic "map."

Image histogram   A graphical representation of the brightness values that comprise an image, with the brightness values displayed along the x-axis of the graph and the frequency of occurrence of each of the brightness values on the y-axis.

Linear contrast stretch   Linearly expands the original brightness values of the image to the full range of sensitivity of the display device.

Low-pass filters   Filters designed to emphasize larger, homogeneous areas of similar brightness values and suppress local details in an image. In other words, low-pass filters are used to smooth (or to blur) the appearance of an image.

Maximum likelihood classifier   This algorithm assumes that the training data statistics for each class in each band are normally distributed, that is, the Gaussian distribution. Under this assumption, the mean vector and the covariance matrix can measure the distribution of spectral response pattern of a class. A probability density function value is calculated for each class according to these parameters. To classify an unknown pixel, the probability value of the pixel belonging to each class is calculated, and it is assigned to the class that has the largest (or maximum) value.

Minimum distance classifier   Computes the mean vectors of all bands for each class based on the training data sets. An unknown pixel is classified by computing the distance between the DN value of this unknown pixel and each mean vector. The unknown pixel is then assigned to a class to which it has the minimum distance in the spectral space.

Nearest neighbor resampling   Uses the input cell value closest to the output cell as the assigned value to the output cell.

Non-systematic errors of satellite images   Errors caused by the sensor platform departing from its normal altitude or the change in terrain elevation, and by the failure of the axis of a sensor system to maintain normal to the Earth's surface and the other parallel to the platform's flight direction.

Object-oriented classification   Two steps are involved in the classification process. Image segmentation merges pixels into objects, and a classification algorithm is then applied to the objects instead of individual pixels.

Parallelepiped classifier   Calculates the range of DN values in each category of training data sets. An unknown pixel is assigned to a class in which it lies.

Principal component analysis   A mathematical technique that transforms correlated variables (e.g., image bands) to fewer principal components uncorrelated to each other.

Radiometric correction   Correction of errors or inconsistencies in image brightness value, that is, radiometric noise.

Radiometric noise   Errors or inconsistencies in image brightness values occurring in raw images due to variations in scene illumination, viewing geometry, atmospheric conditions, or sensor noise and response characteristics. These errors or inconsistencies vary with the type of sensor and platform used to acquire the data and the conditions during data acquisition.

Rectification   The process by which the geometry of an image is made planimetric.

Registration   The process of geometrically aligning two or more sets of image data such that resolution cells for a single ground area can be digitally or visually superimposed.

Relative radiometric correction   An image preprocessing procedure used to normalize the intensities among different bands within an image or to normalize the intensities of bands of one date to a standard scene chosen by the analyst. The correction can be done by single-image adjustment using histogram adjustment, or by multiple-data image normalization using regression.

Resampling　The process of extrapolating data values to a new grid. It is the step in rectifying an image that calculates pixel values for the rectified grid from the original data grid.

Spatial enhancement　A group of image enhancement techniques that emphasize the spatial relationship of image pixels. These techniques explore the distribution of pixels of varying brightness over an image by defining homogeneous regions and by detecting and sharpening boundary discontinuities between adjacent pixels with notable differences in brightness values.

Spatial filtering　The method that uses user-defined filters to emphasize or to suppress image data of different spatial frequencies.

Spatial frequency　The manner in which brightness values change relative to their neighbors in an image.

Spectral angle mapper　A commonly used method for comparing an image spectrum to a reference spectrum that treats both spectra as vectors in a multidimensional space and the number of dimensions equals the number of spectral bands. The spectral angle between the two spectra is computed for determination of matching.

Spectral mixture analysis　It assumes that mixed spectral signatures are caused by the mixture of a set of "pure" materials, called end members, whose spectra can be measured in the laboratory, in the field, or from the image itself. A spectral mixing model is developed to invert the proportions of those "pure materials."

Supervised classification　The analyst identifies in the imagery homogeneous representative samples of different cover types of interest to be used as training areas. The spectral signature will then be determined for the class based on the DNs of training areas. Each pixel in the imagery would compare spectrally with these signatures to determine to which information class it should belong.

Systematic errors of satellite images　Errors caused by scan skew, mirror-scan velocity variance, platform velocity, Earth rotation, and the curvature of the Earth's surface.

True or natural color composite　A true/natural color image may be generated by projecting blue, green, and red spectral bands through the blue, green, and red color guns, respectively. Colors in this type of image composite resemble a visible color photograph, that is, vegetation in green, water in blue, soil in brown, and so on.

Unsupervised classification　Spectral classes are first grouped based solely on digital numbers of the band using a statistical method. Later, these spectral groups can be combined, disregarded, or broken down into smaller groups. The identities of the spectral groups need to match with information classes by using some forms of references (e.g., a thematic map or aerial photographs).

Review Questions

1.   Every satellite image is obtained in black-and-white at a precise wavelength. Why do some satellite images have different colors when seen on a computer screen?

2.   Examine the Landsat TM color composite images of Terre Haute, Indiana, U.S.A. (Fig. 8.15). Identify one or two uses for each color composite.

3.   Radiometric correction is not necessarily equally useful for all remote sensing applications. Can you illuminate on a situation where absolution radiometric correction should be conducted and another situation when relative radiometric correction is necessary?

4.   For your particular application (forestry, land cover, geology, and so forth) of Landsat TM images, which of the contrast enhancement methods discussed is optimum? Explain the reasoning for your selections and any trade-offs that might occur.

5.   Explain the principle that Fourier Transform is used for noise removal and other appropriate image processing tasks.

6.   Why have so many different vegetation indices been developed? Why not use just one? A vegetation index is not necessarily equally useful for all study areas. For a subject of interest to you (land cover, forestry, geology, etc.) evaluate the significance of the use of vegetation indices by citing examples of how these indices might be used. Also, list some situations for which vegetation indices might be less useful.

7.   For land cover analysis, atmospheric correction is not as important as that for a hydrologic or geologic analysis. Justify this statement with reasons.

8.   In conducting geometric correction, is the distributional pattern of ground control points important? Why is it sometimes necessary to select questionable points in order to obtain a geographically even distribution?

9.   Compare and contrast between supervised and unsupervised image classification in terms of advantages and disadvantages. When would one approach be more appropriate than the other?

10.   Evaluate the usefulness of image classification in analysis of land cover and forestry.

11.   What are some possible effects of very high spatial resolution satellite imagery on image classification? (Hint: consider object-based image classification.)

12. Explain why and how hyperspectral images need to have image processing and information extraction methods different from those for multispectral imagery.

References

Arnold, R. H. 2004. *Interpretation of Airphotos and Remotely Sensed Imagery*. Long Grove, Ill.: Waveland Press, Inc.

Avery, T. E., and G. L. Berlin. 1992. *Fundamentals of Remote Sensing and Airphoto Interpretation*, 5th ed. Upper Saddle River, N.J.: Prentice Hall.

Benediktsson, J. A., J. R. Sveinsson, and K. Arnason. 1995. Classification and feature-extraction of AVIRIS data. *IEEE Transactions on Geoscience and Remote Sensing*, 33(5):1194–1205.

Canada Centre for Remote Sensing. 2007. *Tutorial: Fundamentals of Remote Sensing*, http://www.ccrs.nrcan.gc.ca/resource/tutor/fundam/chapter2/08_e.php. Accessed on December 3, 2010.

Campbell, J. B. 2007. *Introduction to Remote Sensing*, 4th ed. New York: The Guilford Press.

Eckhardt, D. W., J. P. Verdin, and G. R. Lyford. 1990. Automated update of an irrigated lands GIS using SPOT HRV imagery, *Photogrammetric Engineering & Remote Sensing*, 56(11):1515–1522.

Filippi, A.M., and J. R. Jensen. 2006. Fuzzy learning vector quantization for hyperspectral coastal vegetation classification. *Remote Sensing of Environment*, 100(4):512–530.

Gitas, I. Z., G. H. Mitri, and G. Ventura. 2004. Object-based image classification for burned area mapping of Creus Cape Spain, using NOAA-AVHRR imagery. *Remote Sensing of Environment*, 92:409–413.

Goetz, A. F. H., G. Vane, J. E. Solomon, and B. N. Rock. 1985. Imaging spectrometry for earth remote sensing. *Science*, 228(4704):1147–1153.

Gong, P., R. Pu, G. S. Biging, and M. Larrieu. 2003. Estimation of forest leaf area index using vegetation indices derived from Hyperion hyperspectral data. *IEEE Transactions on Geoscience and Remote Sensing*, 41(6):1355–1362.

Green, A. A., M. Berman, P. Switzer, and M. D. Craig. 1988. A transformation for ordering multispectral data in terms of image quality with implications for noise removal. *IEEE Transactions on Geoscience and Remote Sensing*, 26:65–74.

Jensen, J. R. 2005. *Introductory Digital Image Processing: A Remote Sensing Perspective*, 3rd ed. Upper Saddle River, N.J.: Pearson Prentice Hall.

Kruse, F. A., A. B. Lefkoff, and J. B. Dietz. 1993. Expert system-based mineral mapping in northern Death Valley, California/Nevada, using the airborne visible/infrared imaging spectrometer (AVIRIS). *Remote Sensing of Environment*, 44:309–336.

Liew, S. C. 2001. *Principles of Remote Sensing*. A tutorial as part of the "Space View of Asia, 2nd ed." (CD-ROM), the Centre for Remote Imaging, Sensing and Processing (CRISP), the National University of Singapore. http://www.crisp.nus.edu.sg/~research/tutorial/rsmain.htm. Accessed January 25, 2011.

Lillesand, T. M., R. W. Kiefer, and J. W. Chipman. 2008. *Remote Sensing and Image Interpretation*, 6th ed. Hoboken, NJ: John Wiley & Sons.

Lo, C. P., and A. K. W. Yeung. 2002. *Concepts and Techniques of Geographic Information Systems*. Upper Saddle River, N.J.: Prentice Hall.

Lu, D., and Q. Weng. 2007. A survey of image classification methods and techniques for improving classification performance. *International Journal of Remote Sensing*, 28(5):823–870.

Price, K. P., X. Guo, and J. M. Stiles. 2002. Optimal Landsat TM band combinations and vegetation indices for discrimination of six grassland types in eastern Kansas. *International Journal of Remote Sensing*, 23:5031–5042.

Pringle, M. J., M. Schmidt, and J. S. Muir. 2009. Geostatistical interpolation of SLC-off Landsat ETM plus images. *ISPRS Journal of Photogrammetry and Remote Sensing*, 64: 654–664.

Pu, R., and P. Gong. 2011. Remote sensing of ecosystem structure and function. In Weng, Q., Ed. *Advances in Environmental Remote Sensing: Sensors, Algorithms, and Applications*. Boca Raton, Fla.: CRC Press, 101–142.

Rajasekar, U., and Q. Weng. 2009. Urban heat island monitoring and analysis by data mining of MODIS imageries. *ISPRS Journal of Photogrammetry and Remote Sensing*, 64(1):86–96.

Roy, D. P., J. Ju, P. Lewis, C. Schaaf, F. Gao, M. Hansen et al. 2008. Multi-temporal MODIS-Landsat data fusion for relative radiometric normalization, gap filling, and prediction of Landsat data. *Remote Sensing of Environment*, 112:3112–3130.

Sabins, F. F. 1997. *Remote Sensing: Principles and Interpretation*, 3rd ed. New York: W.H. Freeman and Company.

Short, N. M. Sr. 2009. *The Remote Sensing Tutorial*, http://rst.gsfc.nasa.gov/. Accessed January 25, 2011.

Song, C., C. E. Woodcock, K. C. Seto, M. P. Lenney, and S. A. Macomber. 2001. Classification and change detection using Landsat TM data: when and how to correct atmospheric effect. *Remote Sensing of Environment*, 75:230–244.

Tong, X., S. Liu, and Q. Weng. 2009. Geometric processing of Quickbird stereo imagery for urban land use mapping— a case study in Shanghai, China. *IEEE Journal of Selected Topics in Applied Earth Observations & Remote Sensing*, 2(2):61–66.

USGS. 2004. Phase 2 gap-fill algorithm: SLC-off gap-filled products gap-fill algorithm methodology, http://landsat.usgs.gov/documents/L7SLCGapFilledMethod.pdf. Accessed January 14, 2011.

Walter, V. 2004. Object-based classification of remote sensing data for change detection. *ISPRS Journal of Photogrammetry & Remote Sensing*, 58:225–238.

CHAPTER **9**

Thermal Remote Sensing

9.1 Introduction

Thermal-IR remote sensing has a long history. Its early development was closely associated with military use. It was not until the mid-1960s when a few manufacturers were permitted to acquire thermal images capable of identifying geologic and terrain features (Sabins, 1997). Satellite acquisition of thermal-IR images began as early as 1964 with the first weather satellite, TIROS-1, but its coarse spatial resolution was suitable only for analysis of cloud patterns, ocean temperatures, and very large terrain features. In 1978, NASA launched an experimental satellite program, the Heat Capacity Mapping Mission (HCMM), which was dedicated to determining how useful temperature measurements could be in identifying materials and thermal conditions on the land and sea, with 600 m spatial resolution for its thermal-IR band and temperature resolution of 0.4 K. Landsat-4 successfully carried a thermal-IR sensor into space with spatial resolution of 120 m. In the meantime, NASA also developed a few airborne thermal-IR sensors, such as the thermal infrared multiple scanner (TIMS) in 1980. In addition, the public may have been aware of ground studies with thermal-IR instruments, such as portable or hand-held thermal imaging equipment or cameras, most of which can operate at night. These instruments have been widely used in such fields as firefighting, heat loss from houses, medical treatment, and military combat. In this chapter, we first discuss the thermal radiation properties and processes, and then move onto the characteristics of thermal images and collection. We end our discussion with examples of thermal image interpretation and analysis in various fields. Since thermal–IR sensors were discussed in Chap. 6, we will not be repeating the discussion here.

9.2   Thermal Radiation Properties

Kinetic heat is the energy of particles in matter in random motion. When matter has a temperature higher than absolute zero ($-273°C$, or 0 K), all particles start their motion. The particles collide, leading to changes in energy state and emission of electromagnetic radiation from the surfaces of materials. Temperature is a measure of the concentration of the kinetic heat, that is, the thermal state of an object. Temperature can be measured by using some sort of instrument. By using an inserted thermometer, we measure the internal/kinetic temperature (resulted from the kinetic motion) of an object (T_{kin}). The electromagnetic energy radiated from an object is the radiant flux, which is often measured in watts per square centimeter ($W \cdot cm^{-2}$). The radiant flux is the external manifestation of an object's thermal state. The radiant temperature (T_{rad}) may be measured by remote sensing instruments known as radiometers (see Sec. 6.5 in Chap. 6 for a detailed description). The kinetic temperature and radiant temperature are highly positively correlated. This linkage makes it possible to use radiometers to measure kinetic temperatures of the Earth's surface materials. However, radiant temperature is always a bit less than kinetic temperature owing to a thermal property called emissivity (to be discussed later).

Kinetic heat may be transferred from one to another location by conduction, convection, or radiation (Sabins, 1997). Heat conduction is the transfer of thermal energy between regions of matter due to a temperature gradient via molecular contacts. Heat spontaneously flows from a region of higher temperature to a region of lower temperature to reduce the temperature difference over time. For example, a small amount of heat is conducted to the Earth's surface constantly from its interior. Convection transfers heat from one place to another by the movement of fluids, and is usually the dominant form of heat transfer in liquids and gases. Hot springs and volcanoes transfer heat by convection. Radiation transfers heat in the form of electromagnetic waves. As discussed in Chap. 2, solar energy reaches the Earth via radiation.

9.3   Thermal Infrared Atmospheric Windows

Although radiant energy from the Earth's surface, as detected by thermal-IR sensors, may come from radiation (solar insolation and sky radiance), convection (atmospheric circulation), and conduction (through the ground), most radiant energy has its origin in solar illumination as well as cloud cover. Heat transfers into and out of near surface layers by the thermal processes of conduction, convection, and radiation. Some of this solar energy travels through the atmosphere, and is absorbed by the Earth's surface materials when it

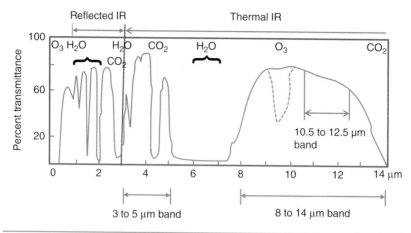

FIGURE 9.1   Atmospheric windows and absorption bands in the thermal-IR spectrum.

arrives. This energy is later emitted from the Earth's surface, but at much longer wavelengths than 0.58 µm. Before this energy reaches thermal-IR sensors, either airborne or spaceborne, it has to go through the atmosphere again. However, the atmosphere does not transmit all wavelengths of the thermal-IR energy equally. In certain regions of the thermal-IR spectrum (3 to 5 µm and 8 to 14 µm), the atmosphere is relatively transparent, allowing the IR energy to reach the sensors. These regions are called atmospheric windows ("peaks" in Fig. 9.1) with which thermal remote sensing is possible. At other regions, carbon dioxide (CO_2), ozone (O_3), and water vapor (H_2O) strongly absorb the thermal-IR energy, making it impossible to observe the Earth's surface from space. These regions are known as absorption bands, as indicated by "valleys" in Fig. 9.1. The spectrum of 5 to 7 µm is an absorption band where water vapor is the dominant absorber of the IR energy. The ozone layer at the top of the atmosphere absorbs much of the energy from approximately 9.2 to 10.2 µm (indicated by a dashed line in Fig. 9.1). To avoid this absorption band, satellite thermal remote sensing utilizes only the spectrum from 10.5 to 12.5 um. Because aircrafts fly underneath the ozone layer, thermal sensors onboard the aircrafts are not affected by this absorption band, and can thus measure the energy from the full window of 8 to 14 µm (Sabins, 1997).

9.4   Thermal Radiation Laws

We have to know works by a few physics giants in order to understand the principles of thermal-IR remote sensing. Russian scientist Gustav Robert Kirchhoff invented the concept of a "blackbody" radiation in 1862. Blackbody is a hypothetical, ideal radiator that totally absorbs and reemits all energy incident upon it. If a radiation-emitting object meets the physical characteristics of a blackbody in thermodynamic equilibrium, the radiation is called "blackbody radiation."

The emitted frequency spectrum of the blackbody radiation is described by a probability distribution depending only on temperature given by Planck's law of blackbody radiation. In other words, Planck's law describes the spectral radiance of electromagnetic radiation at all wavelengths emitted from a blackbody at a specific absolute temperature. The amount of the radiant energy from a blackbody depends upon two parameters: temperature and wavelength. Wien's displacement law further provides the most likely frequency of the emitted radiation, while the Stefan-Boltzmann law determines the radiant intensity.

Figure 9.2 shows the spectral distribution of the energy radiated from the surface of a blackbody at various temperatures. It is interesting to observe changes in the radiant energy peak, the

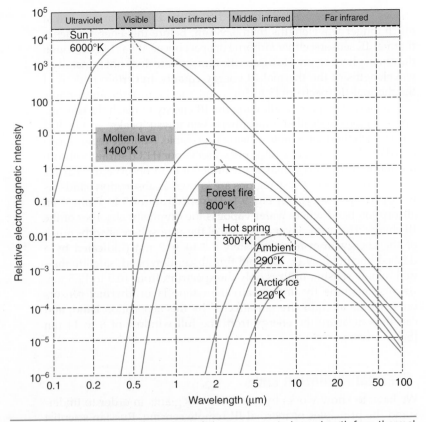

FIGURE 9.2 Emitted radiance (as intensity) versus spectral wavelength from thermal radiators at various peak radiant temperatures (*After Short, 2009*).

wavelength at which the maximum amount of energy is radiated. As temperature increases, the radiant energy peak shifts toward shorter wavelengths. This shift (displacement) in the radiant energy peak with increased temperature for a blackbody is explained by the Wien Displacement Law:

$$\lambda_m = 2898/T_{rad} \qquad (9.1)$$

where λ_m is the wavelength at maximum radiant exitance and T_{rad} is the radiant temperature expressed in Kelvin. The physical constant, 2898, is in units of $\mu m \cdot K$. If we substitute the sun's radiant temperature of about 6000 K into Eq. (9.1), its radiant energy peak is obtained, which is close to 0.58 μm (in the visible spectrum). The Earth's surface has an average temperature of approximately 300 K, and therefore its peak emission occurs at approximately 9.7 μm. This peak emission falls within the 8- to 14-μm interval, an atmospheric window. For this reason, thermal remote sensing of most of the Earth's surface features is appropriate by using this spectral interval. However, a forest fire possesses a much higher temperature (e.g., 800 K), as does volcano lava (e.g., 1400 K). Their peak emissions will be approximately 3.62 μm and 2.07 μm, respectively. Remote sensing of these particular features will be suitable for using the other atmospheric window with spectral range of 3 to 5 μm.

The total amount of radiant energy coming from the surface of a blackbody (the area underneath each curve in Fig. 9.2) increases as temperature increases. Therefore, if a sensor can measure spectral radiant exitance across all wavelengths, the total radiant energy from the blackbody at any given temperature will be proportional to the area under the radiation curve. The hotter the radiating body, the greater its radiance over its range of wavelengths will be. This relationship between the total radiant exitance and the spectral distribution pattern of a blackbody is further described in the Stefan-Boltzmann Law, which states that:

$$M_b = \sigma T_{kin}^{4} \qquad (9.2)$$

where M_b is the total radiant exitance of a blackbody in Watts per square meter, and T_{kin} is its kinetic temperature in Kelvin units. The Stefan-Boltzmann constant, σ, is given as 5.67×10^{-12} W $\times$ cm^{-2} $\times$ K^{-4}. This equation indicates that the total radiant flux from the surface of a blackbody is a function of the fourth power of its absolute temperature. In thermal remote sensing, the radiant exitance from a surface is used to infer the temperature of that surface.

The concept of a blackbody is convenient for scientific thinking, but real materials do not behave as a blackbody. In reality, no material absorbs all energy incident upon it and radiates the full amount of energy given in Eq. (9.2). All real materials emit only a fraction of the energy emitted from a blackbody at the equivalent temperature.

Emissivity (ε) is a coefficient that describes how efficiently an object radiates energy compared to a blackbody, and is given as follows:

$$\varepsilon = \frac{M_r}{M_b} \qquad (9.3)$$

where M_r is the radiant exitance of a real object at a given temperature and M_b is the radiant exitance of a blackbody at the same temperature. All real materials have an ε value between 0 and 1, which is spectrally dependent, that is, it changes with wavelength. In addition, water content, chemical composition, structure, and roughness also influence the emissivity of ground features (Snyder, Wan, Zhang, and Feng, 1998). Table 9.1 lists the emissivities of common materials in the wavelength region of 8 to 14 μm.

The relationship between emissivity and absorptivity is manifested by Kirchoff's radiation law. It states that at thermal equilibrium, the spectral emissivity of an object (or surface) equals its absorptance. In other words, a good absorber is a good emitter. In Chap. 2, we discussed the principle of energy conservation, and established the relationship between the incident radiation and three types of energy interactions with a terrain feature, that is, reflected, absorbed, and transmitted radiations. Because real materials are usually opaque to thermal-IR radiation, we can modify Eq. (2.3) by dropping the term of transmittance and establishing a new equation of energy balance:

$$\text{Reflectance} + \text{Absorptance} = 1 \qquad (9.4)$$

Substituting absorptance with emissivity results in

$$\text{Reflectance} + \text{Emissivity} = 1 \qquad (9.5)$$

Equation (9.5) indicates that the higher an object's reflectance, the lower its emissivity, and vice versa. In other words, a good reflector is a poor emitter. For example, granitic rock has a higher reflectance than wet soil; therefore, its emissivity (approximately 0.85) is substantially lower than that of wet soil (0.95 to 0.98). Similarly, dry snow (0.85 to 0.90) has a lower emissivity than wet snow (0.98 to 0.99).

The measurement of radiant temperature from remotely sensed thermal-IR data requires an accurate measurement of emissivity value of the surface (Weng, Lu, and Schubring, 2004). Based on the Stefan-Boltzmann Law and our discussion about emissivity, we can calculate the radiant flux of a real material as follows:

$$M_r = \varepsilon \sigma T_{kin}^{\ 4} \qquad (9.6)$$

where ε is the emissivity of that material. From this equation, it is clear that two materials of the same kinetic temperatures may yield

Material	Emissity (ε)	Material	Emissity (ε)
Highly polished gold	0.02–0.03	Granite, typical	0.815
Aluminum foil	0.03–0.07	Dunite	0.856
Polised metals	0.16–0.21	Obsidian	0.862
Sheet iron (rusted)	0.63–0.70	Feldspar	0.87
Glass	0.77–0.81	Granite, rough	0.898
Granite, typical	0.815	Silicon sandstone, polished	0.909
Granitic rock	0.83–0.87	Sand, quartz, large-grain	0.914
Dry snow	0.85–0.90	Dolomite, polished	0.929
Dunite	0.856	Basalt, rough	0.934
Obsidian	0.862	Dolomite, rough	0.958
Feldspar	0.87	Asphalt paving	0.959
Dry vegetation	0.88–0.94	Concrete walkway	0.966
Granite, rough	0.898	Water, with a thin film of petroleum	0.972
Paint	0.90–0.96	Water, pure	0.993
Silicon sandstone, polished	0.909	Highly polished gold	0.02–0.03
Sand, quartz, large-grain	0.914	Aluminum foil	0.03–0.07
Portland cement concrete	0.92–0.94	Polised metals	0.16–0.21
Dry mineral soil	0.92–0.94	Sheet iron (rusted)	0.63–0.70
Basaltic rock	0.92–0.96	Glass	0.77–0.81
Dolomite, polished	0.929	Granitic rock	0.83–0.87
Wood	0.93–0.94	Dry snow	0.85–0.90
Brick	0.93–0.94	Dry vegetation	0.88–0.94
Basalt, rough	0.934	Paint	0.90–0.96
Asphaltic concrete	0.94–0.97	Portland cement concrete	0.92–0.94
Wet soil	0.95–0.98	Dry mineral soil	0.92–0.94
Dolomite, rough	0.958	Basaltic rock	0.92–0.96
Asphalt paving	0.959	Wood	0.93–0.94
Healthy green vegetation	0.96–0.99	Brick	0.93–0.94
Concrete walkway	0.966	Asphaltic concrete	0.94–0.97
Rough ice	0.97–0.98	Wet soil	0.95–0.98
Human skin	0.97–0.99	Healthy green vegetation	0.96–0.99
Water, with a thin film of petroleum	0.972	Rough ice	0.97–0.98
Wet snow	0.98–0.99	Human skin	0.97–0.99
Clear water	0.98–0.99	Wet snow	0.98–0.99
Water, pure	0.993	Clear water	0.98–0.99

Sources: Lillesand et al. 2008; Sabins, 1997.

TABLE 9.1 Emissivities of Common Materials

different radiant exitance due to the difference in spectral emissivity. Radiant temperature is a more favorite parameter in thermal remote sensing because the thermal-IR sensors measure radiant temperature instead of radiant flux. The radiant temperature is related to kinetic temperature as follows:

$$T_{rad} = \varepsilon^{1/4} T_{kin} \qquad (9.7)$$

From this equation, we know that radiant temperature is always lower than kinetic temperature. However, in a material that has an emissivity close to one, such as clear water, the radiant temperature may be assumed equal to the kinetic temperature. In most cases, correction for emissivity is necessary in estimating kinetic temperatures from remotely sensed radiant temperatures. Lack of knowledge of emissivity can introduce an error ranging from 0.2 K to 1.2 K for mid-latitude summers and from 0.8 K to 1.4 K for winter conditions for an emissivity of 0.98 at a ground height of 0 km, when a single channel of thermal-IR data is used for estimating the kinetic temperatures (Dash, Gottsche, Olesen, and Fischer, 2002).

9.5 Thermal Properties of Terrestrial Features

Remotely sensed temperature measurements and thermal responses often provide important information about the nature of the composition and other physical attributes of materials at the Earth's surface. For any material, three thermal properties play important roles in governing the temperature of a material at equilibrium and its temporal dynamics. These properties include the material's thermal conductivity, capacity, and inertia. Thermal conductivity is a measure of the rate at which heat passes through a material of a specific thickness. It is expressed as the calories delivered in 1 sec across a 1-cm^2 area through a thickness of 1 cm at a temperature gradient of 1°C (unit: $cal \cdot cm^{-1} \cdot sec^{-1} \cdot °C$). Thermal capacity denotes the ability of a material to store heat. It is a measure of the increase in thermal energy content per degree of temperature rise, specifically the number of calories required to raise the temperature by 1°C (unit: $cal \cdot °C^{-1}$). A related measure is specific heat, which refers to the heat capacity of 1 g of a material. Thermal inertia is a measure of the resistance of a material to temperature change, and is expressed as calories per square centimeter per second square root per degree Celsius (unit: $cal \cdot cm^{-1} \cdot sec^{-1/2} \cdot °C^{-1}$). Thermal inertia (P) is defined as:

$$P = (Kc\rho)^{1/2} \qquad (9.8)$$

where K is thermal conductivity, ρ is density, and c is thermal capacity. The density of a material is very important, as thermal inertia usually increases with increasing density (Sabins, 1997). Thermal inertia is indicated by the time-dependent variations in temperature during a 24-hour heating/cooling cycle. The materials with high P possess a strong inertial resistance to temperature fluctuations, and they show less temperature variation during the heating/cooling cycle. In contrast, the materials with low P have relatively larger differences between maximum and minimum temperature during the heating/cooling cycle.

Figure 9.3 shows daily (diurnal) variations of radiant temperatures for some typical terrestrial features. During a diurnal cycle, the near surface layers of the Earth experience alternate heating and cooling to depths typically ranging from 50 to 100 cm, or 20 to 40 in. Solar radiation and heat transfer from the air significantly heat up materials at and immediately below the surface during the day, while temperatures usually drop at night primarily by radiative cooling, accompanied by some conduction and convection (Short, 2009). In addition, seasonal changes in temperature and local meteorological conditions also affect the diurnal temperature cycle. Rocks and dry soils experience larger temperature variations, relating to their low thermal inertia (Larson and Carnahan, 1997). Moist soils observe much less temperature variation than dry soils because moist soils contain water and tend to have higher density. The water content increases the thermal capacity, while increased density enhance their thermal inertia. Furthermore, the emissivity of soils influences radiant temperatures as well, which is a function of soil moisture and soil density (Larson and Carnahan, 1997). Water has a smaller temperature variation. This is partly due to its rather high thermal inertia, relative to typical land surfaces, as controlled largely by water's high specific heat. Vegetation tends to have a bit higher temperature than water throughout the day, but a bit lower temperature at nighttime. In order to understand fully the thermal

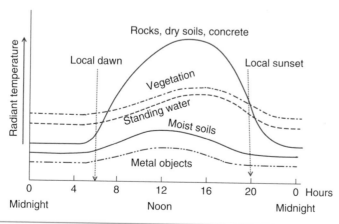

FIGURE 9.3   Daily (diurnal) variations of radiant temperatures for some typical terrestrial features (*After Sabins, 1997*).

response characteristics of vegetation, it is often necessary to scrutinize the temperatures of each part of the vegetation-ground system (such as shaded ground, sunny ground, shaded vegetation, and sunny vegetation) and to examine the effects of different canopy structures (Kimes, 1983; Cassels, Sobrino, and Coll, 1992a,b). Thermal responses for vegetation can be highly varied as a function of the biophysical properties of the vegetation itself (Quattrochi and Ridd, 1998). Concrete and asphalt surfaces absorb more heat and remain warmer throughout the night than other terrestrial features, but they typically have high diurnal variations in temperature. Metal objects are generally good reflectors, so their emissivities are low. This explains why they keep low temperature at both daytime and nighttime.

A closer look at Fig. 9.3 shows that most rapid changes in temperatures occur near dawn and sunset for all materials. Further, we find that water reaches its peak temperature an hour or two after the other materials. However, the diurnal curves for all materials intersect shortly after dawn and near sunset. These intersections are called "thermal crossovers," the time when radiant temperatures of these materials are the same. Geologists prefer to collect predawn thermal imagery for their work because during this time most terrestrial materials have relatively stable temperatures. The shadows caused by shaded trees, buildings, and other features, often existing in daytime thermal imagery, do not occur in the predawn imagery (Jensen, 2007; Lillesand, Kiefer, and Chipman, 2008). Various temperature heating and cooling rates and temporal dynamics not only can help for the selection of times for thermal image acquisition, but also can help greatly in the interpretation of thermal-IR imagery.

9.6   Characteristics of Thermal Images and Collection

The collections of thermal-IR images need to consider both spatial resolution and temperature resolution. For an across-track scanning system, the spatial resolution may be different in the azimuthal (flight) direction and in the scan direction. Spatial resolution is a function of the IFOV of the scanner, the altitude of the remote sensing platform, and the scan angle. The farther a pixel is from the nadir, the coarser the spatial resolution is. The temperature resolution of a thermal-IR scanner manifests how accurate and how fine temperature from the terrain can be measured. The accuracy of radiant temperature measurements for current thermal-IR sensors can reach 0.1°C. Spatial resolution and temperature resolution have an inverse relationship. To improve temperature resolution, a larger IFOV is needed so that more thermal-IR energy can be received. The temperature resolution of thermal-IR images is equivalent to the radiometric resolution for reflected bands. On thermal-IR images, bright tones indicate warm temperatures, while dark tones correspond to cold areas. The DNs of thermal-IR

images can be converted to radiant temperatures through some computational procedures (Sobrino, Jiménez-Muñoz, and Paolini, 2004;
Weng et al., 2004). Some projects do not require having accurate measures of ground temperatures, and therefore relative temperature
differences among various materials/objects are sufficient. When
absolute ground temperatures are needed, the computational procedures are more complicated in order to separate temperatures and
emissivities as well as to correct the atmospheric effects.

Thermal-IR sensors have been flown on both satellites and aircrafts. The earliest satellite, Heat Capacity Mapping Mission (HCMM)
launched on April 26, 1978 by NASA, carried a thermal sensor system
capable of measuring temperature, albedo, and thermal inertia. The
satellite passed the equator at approximately 2:00 PM as well as 1:30
and 2:30 AM local time. HCMM used a thermal radiometer to detect
emitted thermal-IR radiation in the 10.5- to 12.5-µm interval with spatial resolution of 600 m, but it did not last for long. Presently, only a
few sensors have thermal-IR capabilities of global imaging. The Landsat TM sensor onboard Landsat-5 has been acquiring thermal-IR
images of the Earth nearly continuously from July 1982 to the present,
with a single thermal-IR band of 120-m resolution, and is thus long
overdue. Another thermal-IR sensor that has global imaging capacity
is with Landsat-7 ETM+. Again, it is a single thermal-IR band, but the
spatial resolution is 60 m. As discussed in Chap. 7, NOAA's AVHRR
has two thermal-IR channels, with wavelengths from 10.30 to 11.30 µm
and 11.50 to 12.50 µm, respectively, but provides data with a much
coarser spatial resolution at 1100 m. AVHRR thermal-IR images have
been widely employed for cloud and surface temperature mapping at
daytime and nighttime. Terra's ASTER multispectral thermal-IR bands
of 90-m resolution have been increasingly used in urban climate and
environmental studies in recent years (Weng, 2009). Table 9.2 outlines
the specific characteristics of each thermal-IR band. ASTER is an on-
demand instrument, which means that data are only acquired over the
requested locations. Figure 9.4 shows an example using ASTER thermal-IR data to monitor power plants. At the top, an ASTER false color

Band	Label	Wavelength (µm)	Spatial Resolution (m)
B11	TIR_Band10	8.125–8.475	90
B12	TIR_Band11	8.475–8.825	90
B13	TIR_Band12	8.925–9.275	90
B14	TIR_Band13	10.250–10.950	90
B15	TIR_Band14	10.950–11.650	90

TABLE 9.2   Terra's ASTER Thermal-IR Bands

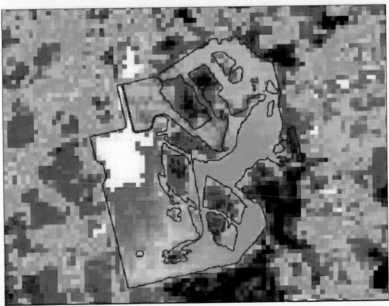

FIGURE 9.4   Discharge heated effluent water from a power plant as shown in an ASTER thermal-IR image. Its visible and near-IR (15 m spatial resolution) is used to create a color composite image (top), which shows the power plant in bright blue-white. Its thermal-IR image (90 m resolution) is used to derive a surface temperature map (bottom). (*Images courtesy of NASA/GSFC/MITI/ERSDAC/JAROS, and U.S./Japan ASTER Science Team.*)

image was created by using its visible and near-IR bands to display Braidwood nuclear power plant in Illinois, located approximately 75 km (47 mi) southwest of Chicago, Illinois. Braidwood Nuclear can be seen in the top image as the bright blue-white pixels just above the large cooling pond. Like many power plants, Braidwood Nuclear uses a cooling pond to discharge heated effluent water. In the bottom image, a single ASTER thermal-IR band was color-coded to represent land surface temperature. The warmest areas are shown in white, and progressively cooler areas are in red, orange, yellow, green, and blue. Black corresponds to the coolest areas. We can see the bright white plume of hot water discharged from the power plant. The water grows gradually cooler as it circulates around the pond. Terra satellite launched in December 1999 as part of NASA's EOS has a life expectancy of six years, and is now overdue. In addition, the MODIS sensors onboard Terra and Aqua provide multiple thermal-IR bands. Twenty-nine spectral bands (bands 8 to 36) have 1000-m spatial resolution, and are located in the mid- and thermal-IR regions. The Terra satellite was launched in 1999 as an EOS AM satellite, and the Aqua in 2002 as an EOS PM satellite. Together the MODIS instruments capture thermal-IR data in both daytime and nighttime, which can image the entire Earth every 1 to 2 days.

Two airborne sensors can be used to collect both daytime and nighttime thermal images. TIMS, operated for NASA by JPL, uses the across-track scanning technology. TIMS breaks the 8- to 12-µm interval into six bands with spatial resolution of 18 m: (1) 8.2–8.6 µm, (2) 8.6–9.0 µm, (3) 9.0–9.4 µm, (4) 9.4–10.2 µm, (5) 10.2–11.2 µm, and (6) 11.2–12.2 µm. TIMS data have been used extensively in studies of minerals in rocks and soils and volcanoes in the western United States, the Hawaiian islands, and Europe. Advanced Thermal and Land Applications (ATLAS) is a 15-channel multispectral scanner that basically incorporates the bandwidths of the Landsat TM (along with several additional bands) and six thermal-IR bands similar to the TIMS sensor. ATLAS thermal-IR data, collected at 10-m spatial resolution, have been used to study urban surface energy responses in Huntsville, Alabama (Lo, Quattrochi, and Luvall, 1997).

9.7 Thermal Image Interpretation

Thermal-IR imagery has been applied successfully in many different fields. These fields include identifying the Earth's surface materials (such as types and structures of soils, minerals, and rocks); estimating soil moisture and evapotranspiration from vegetation; monitoring active volcanoes, subsurface fires in landfills and coal mines, and forest fires; determining environmental features such as thermal plumes in lakes and rivers, warm and cold springs, and currents in water bodies; and studying land surface energy balance, land cover, and the urban heat island effect. However, interpreting thermal-IR images and temperature distribution over an area is not straightforward. In some instances, we need to look at the patterns of relative temperature differences, while in other instances we must determine absolute temperature values. Many factors should be taken into account to make quantitative determinations of surface temperatures, especially variations in sun angle, the composition, density, and texture of surface materials, topographic irregularities (elevation, slope, and surface direction relative to the sun's position), rainfall history and moisture variations, and near surface climatic conditions (1 to 3 m air temperature and its history, relative humidity, wind effects, and cloud cover; Short, 2009). Several other environmental factors might be as important under certain circumstances, such as the variations in the sun's position relative to sensor position, emissivities of surface materials, vegetation canopy characteristics (height, leaf geometry and temperature, plant shape, and plant stress), absorption and re-emission of thermal radiation by aerosols, water vapor, and air gases, and contributions from geothermal heat flux (Short, 2009).

The impacts of environmental factors vary from locale to locale, which make it more complicated for the interpretation and analysis of thermal-IR images. The rest of this section illustrates a few examples of thermal image interpretation. The first three examples are adapted from NASA's Earth Observatory, and the other two examples are from NASA's Remote Sensing Tutorials by Nicholas M. Short, Sr. (2009).

Figure 9.5 shows ASTER images of the Saline Valley area, California, acquired on March 30, 2000. Saline Valley is a large, deep, and arid valley in the northern Mojave Desert of California. Most of it became a part of Death Valley National Park when the park expanded in 1994. A number of hot springs exit in the northeast corner of the valley, and the water temperature at the source of these springs reaches 107°F (42°C) on average. Each of the images displays data collected by the ASTER sensor from a different spectral region, and together these images provide comprehensive information about the composition of surface materials in the Saline Valley area. Figure 9.5(*a*) is a false color composite of visible and near-IR bands 3 (0.760–0.860 μm, near-IR band), 2 (0.630–0.690 μm, red band), and 1 (0.520–0.600 μm), displayed in red, green, and blue (RGB) colors. Vegetation appears red, snow and dry salt lakes are white, and exposed rocks are brown, gray, yellow, and blue. Rock colors mainly reflect the presence of iron minerals, and variations in albedo. Figure 9.5(*b*) displays a color composite created by using three short-wavelength IR bands 4 (1.600–1.700 μm), 6 (2.185–2.225 μm), and 8 (2.295–2.365 μm) as RGB. In this wavelength region, clay, carbonate, and sulfate minerals have unique absorption features, resulting in distinct colors in the image. For example, limestone is yellow-green, and purple areas are kaolinite-rich (kaolinite is a clay mineral). Figure 9.5(*c*) is a color composite of

(a) Composite of near-IR, red, and green bands

(b) Composite of shortwave IR bands 4, 6, and 8

(c) Composite of thermal-IR bands of 13, 12, and 10

FIGURE 9.5 Prospecting different types of rocks and soils using ASTER thermal-IR images. Images acquired from the Earth Observatory. (*Image courtesy of NASA, GSFC, MITI, ERSDAC, JAROS, and the U.S./Japan ASTER Science Team.*)

three thermal-IR bands 13 (10.250–10.950 µm), 12 (8.925–9.275 µm), and 10 (8.125–8.475 µm) as RGB. In the thermal-IR region, variations in quartz content appear in more or less red. Carbonate rocks are in green, while mafic volcanic rocks are purple (mafic rocks have high proportions of elements like magnesium and iron). ASTER's ability to identify different types of rock and soil from space using thermal-IR wavelengths of light is one of its unique capabilities.

The Chiliques volcano, situated in northern Chile, has not erupted for over 10,000 years. Geologists had previously considered Chiliques, a simple 5778-m (18,957-ft) stratovolcano with a 500-m (1640-ft) diameter circular summit crater, to be dormant. However, a January 6, 2002, nighttime thermal-IR image from ASTER revealed a hot spot in the summit crater as well as several others along the upper flanks of the volcano's edifice, indicating new volcanic activity. Examination of an earlier nighttime thermal-IR image from May 24, 2000, showed no such hot spots. The pair of ASTER

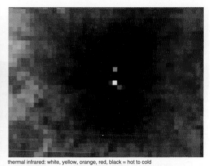

false color: red=near infrared, green=red, blue=green

(a) False color composite image

thermal infrared: white, yellow, orange, red, black = hot to cold

(b) Nighttime thermal-IR image

FIGURE 9.6   A pair of ASTER images for the Chiliques volcano in northern Chile. The daytime image (*a*) was acquired on November 19, 2000, and is displaying as a false color composite using ASTER near-IR, red, and green bands. The nighttime image (*b*) was acquired on January 6, 2002, and is displayed as a color-coded single thermal-IR band. Hot spots can be seen in the thermal-IR image. Images acquired from the Earth Observatory. (*Image courtesy of NASA/GSFC/MITI/ ERSDAC/JAROS, and U.S./Japan ASTER Science Team.*)

images in Fig. 9.6 shows the volcano in a false color composite (*a*) and a nighttime thermal-IR image (*b*). The daytime image of Chiliques was acquired on November 19, 2000, and was created by displaying ASTER bands 3, 2, and 1 in RGB. The nighttime image is a color-coded display of a single thermal-IR band. The hottest areas are white and colder areas are darker shades of red. Both images cover an area of 6.75 × 5 km (4.2 × 3.1 mi), and are centered at 23.6°S latitude, 67.6°W longitude. Hot spots of various temperatures are caused by magma just under the surface of the volcano. Stratovolcanoes, such as Chiliques, account for approximately 60% of Earth's volcanoes. They are marked by eruptions with cooler, stickier types of lava such as andesite, dacite, and rhyolite. Because these types of lava tend to plug up volcanic plumbing, gas pressure can more easily build up to high levels, often resulting in explosive eruptions. The material spewed forth during the eruptions consists roughly of an equal mixture of lava and fragmented rock, so these volcanoes are also commonly known as composite volcanoes.

Figure 9.7 displays a pair of ASTER images, collected on August 27, 2006, to illustrate the contrast between an irrigated poplar plantation and native plants in water use. The poplar plantation to the left of U.S. Highway 84 consumes a large amount of water because poplars are used to grow in wetlands or along riverbanks. To the right of the highway is the Umatilla Ordnance Depot, where native plants, low and desert-hardy, stretch across the flat plain leading to the river. These native plants are suited to hot, dry summers and an average annual precipitation of 200 mm (8 in.), and thus use little water. The top image is an NDVI image, in which dense vegetation appears in

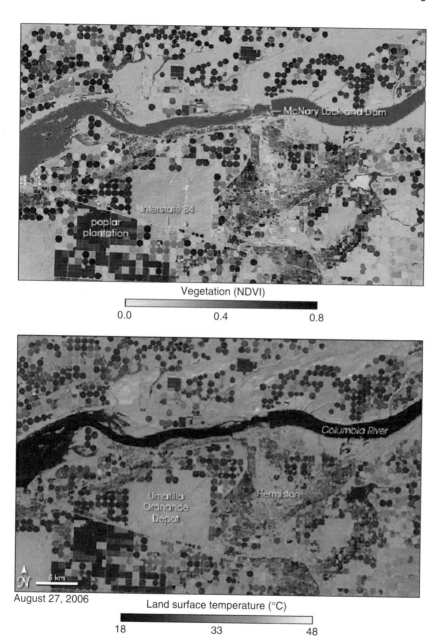

Vegetation (NDVI)

0.0   0.4   0.8

August 27, 2006   Land surface temperature (°C)

18   33   48

FIGURE 9.7   Irrigation and land surface temperature in Oregon, U.S.A. (*NASA image created by Jesse Allen, Earth Observatory, using data provided by NASA/GSFC/ METI/ERSDAC/JAROS, and the U.S./Japan ASTER Science Team.*)

dark green, while sparse vegetation is in pale yellow. Crops watered with pivot irrigation systems form circles in varying shades of green, and the density of the crop growing determines the shade. Freshly harvested fields are a ghostly gold, barely detectable against the surrounding landscape. The native vegetation at the Umatilla Ordnance Depot appears in pale yellow, indicating sparse plants and shrubs with small leaves. The poplar plantation is seen as a dark green square of dense vegetation. The difference between dark green, densely vegetated areas and pale, lightly vegetated areas is water. The regular patterns of green squares and circles indicate that these are irrigated areas. Native, non-irrigated vegetation, by contrast, has an irregular shape. The contrast between irrigated and non-irrigated land can be seen more clearly in the lower image, which shows land surface temperature derived from ASTER thermal-IR data. The coolest areas are in dark blue, while the warmest areas are in pink and yellow. Irrigated croplands are much cooler than the surrounding native vegetation. Thick, leafy plants cool the Earth's surface through evaporation of water, while sparse, small-leafed plants cool the land less effectively. In this semi-arid region, the temperature difference between densely vegetated, irrigated land and lightly vegetated, non-irrigated land can yield as much as 30°C (54°F).

Water is cooler in the daytime and warmer in the nighttime compared with most materials in the Earth. This contrast is due in part to water's rather high thermal inertia, relative to typical land surfaces, which is controlled largely by water's high specific heat (Short, 2009). Therefore, water heats less during the day and holds that heat more at night, giving rise to intrinsically cooler daytime but warmer nighttime temperatures under many meteorological conditions (Short, 2009). Moreover, water in natural settings (rivers, lakes, oceans) is likely to experience thermal convection (e.g., upwelling) and turbulence (e.g., wave action), leading to mixing and temperature homogenization of water at different depths. Water exhibits very dark to medium gray tones in daytime thermal-IR images and moderately light tones in nighttime thermal images. However, the actual appearance of water in the thermal-IR images depends on the time the image is taken, and is controlled by its diurnal temperature pattern. Figure 9.8 illustrates a sequence of four thermal-IR images taken by an airborne Daedalus thermal scanner, which show a land area next to the Delaware River, on the New Jersey side. The images were processed to emphasize the tonal differences in water, and to subdue the land expression of thermal variability. The top three images were taken on the same day, December 28, 1979. The first strip was acquired at 6:00 AM when air temperature was −9°C. Water, appearing in very light tone, was notably warmer (above 0°C) relative to land. Details of the river branches are clearly seen. The second image was obtained at 8:00 AM, when water was still warmer than the land. The bright streak near the land's upper left point is hot

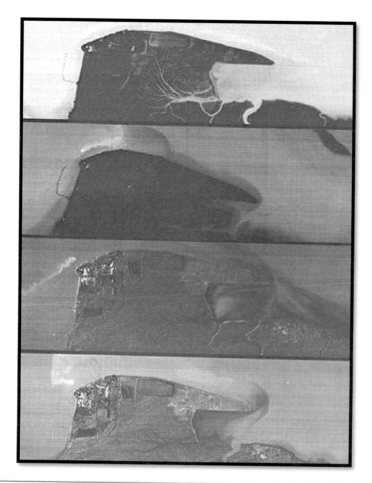

FIGURE 9.8   A sequence of four thermal-IR images taken by an airborne Daedalus thermal scanner on December 28–29, 1979. The images cover a land area next to the Delaware River, on the New Jersey side, in the United States. (*Image acquired from NASA's Remote Sensing Tutorials, http://rst. gsfc.nasa.gov/Sect9/Sect9_5.html.*)

effluent from a power plant. The third image was taken at 2:00 PM, when the air temperature had climbed to –2°C and thermal contrast between land and water was minimal. The power plant, which is a hot building, is clearly visible at the northeast corner of the peninsula. The effluent from the plant is now pointing downstream as ebb tide occurs. The bottom image was taken at 11:00 AM on December 29th, when the air temperature was –4°C. The tide was near high, and the power plant was detailed in the thermal image as well.

The diurnal variations of radiant temperatures are also clearly distinguishable in urban areas. This is evident in Fig. 9.9, with two images

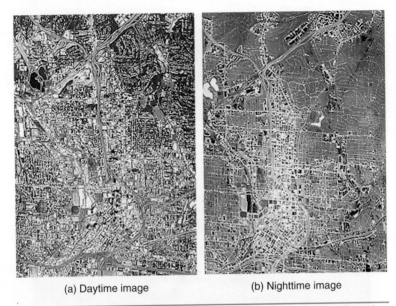

(a) Daytime image   (b) Nighttime image

Figure 9.9   Thermal-IR images of Atlanta Central Business District, Georgia, taken by Advanced Thermal and Land Applications Sensor (ATLAS) on May 11–12, 1997. (*Image courtesy of Dale Quattrochi and Jeff Luvall, Global Hydrology and Climate Center, NASA Marshall Space Flight Center, Huntsville, Alabama.*)

taken over central Atlanta, Georgia, U.S.A. Figure 9.9(*a*) was taken during the day (11 AM to 3:00 PM Eastern Daylight Time) on May 11, 1997, while Fig. 9.9(*b*) was taken just before dawn (2:00 to 4:00 AM) on May 12. The thermal sensor (ATLAS) was flown on a Lear jet aircraft. In the day thermal image, contrasts between heated buildings and streets and areas in shadow create a scene that resembles an aerial photo. Interstate highways 75/85, which traverse in a north-south direction along the city center, are seen as a dark "ribbon" in the day image (Quattrochi and Luvall, 1997). The upper right side of the daytime image has a much darker tone due to low radiant temperature. This low temperature is a consequence of the damping effect that the extensive residential tree canopy has on upwelling thermal energy (Quattrochi and Luvall, 1997). However, in the dawn image, differences in temperature have decreased sharply, and there are no shadows. A part of the image becomes brighter, corresponding to the central business district and nearby areas. Many streets in the city center and residential areas can be clearly identified. The streets, being concrete or asphalt, absorb more heat during the day and remain warmer through the night. The difference in urban temperature and its rural counterpart is known as the "urban heat island" effect, which we detect more easily in the dawn image. Two lakes, seen in the upper left side of the images (look like two cashew nuts), have a strong contrast in tone. The nighttime image shows them in very light tone, while the daytime image is in very dark tone. It appears that thermal energy responses for vegetation across the image are relatively uniform at night, regardless of vegetative type (Quattrochi and Luvall, 1997).

9.8   Summary

All materials on the Earth's surface emit electromagnetic energy if they have an absolute temperature above the freezing point. The amount of radiant energy from a material depends upon two parameters: its kinetic temperature and emissivity. For any material, three thermal properties—thermal conductivity, capacity, and inertia—play important roles in governing the temperature of a material at equilibrium and its change over time. However, the diurnal temperature variation of a specific material is largely controlled by thermal inertia, which has a close relationship with the density of that material.

Thermal-IR images measure radiant temperatures of materials. Due to strong absorption of thermal-IR energy by carbon dioxide, ozone, and water vapor, only certain regions of the thermal-IR spectrum, known as atmospheric windows, can be used for thermal remote sensing. Although thermal images can provide distinctive signatures or indirect indications of properties of materials that are diagnostic, many environmental factors need to be considered as they may influence thermal signatures to different degrees. Thermal-IR sensors often have to compromise between spatial and temperature resolution. Thermal image interpretation and analysis may or may not need to compute accurate ground temperatures, depending on the objective of a specific project. Both spaceborne and airborne thermal-IR images have been applied in various applications, including identifying soils, minerals and rocks, estimating soil moisture and evapotranspiration from vegetation, monitoring active volcanoes and forest fires, determining thermal plumes in lakes and rivers, and studying the urban heat island effect.

Key Concepts and Terms

Blackbody   A hypothetical, ideal radiator that totally absorbs and reemits all energy incident upon it.

Diurnal cycle of radiant temperatures   For some typical terrestrial features, the near surface layers of the Earth experience alternate heating and cooling daily to depths typically ranging from 50 to 100 cm, or 20 to 40 in. Solar radiation and heat transfer from the air significantly heat materials at and immediately below the surface during the day, while temperatures usually drop at night primarily by radiative cooling, accompanied by some conduction and convection.

Emissivity   A coefficient that describes how efficiently an object radiates energy compared to a blackbody.

Heat capacity mapping mission   In 1978, NASA launched an experimental satellite program called the Heat Capacity Mapping Mission. The program was dedicated to determining how useful temperature measurements could be in identifying materials and thermal conditions on land and at sea, with 600 m spatial resolution for its thermal-IR band and temperature resolution of 0.4 K.

Heat conduction   Transfer of thermal energy between regions of matter due to a temperature gradient via molecular contacts.

Heat convection   Transfer of heat from one place to another by the movement of fluids. It is usually the dominant form of heat transfer in liquids and gases.

Kinetic heat   The energy of particles in matter in random motion.

Kinetic temperature   Internal temperature of an object resulting from random motion of particles. It is measured by using an inserted thermometer.

Planck's law   The law describing the spectral radiance of electromagnetic radiation at all wavelengths emitted from a blackbody at a specific absolute temperature. The amount of the radiant energy from a blackbody depends upon two parameters: temperature and wavelength.

Radiant flux   The electromagnetic energy radiated from an object. Often measured in watts per square centimeter ($W \cdot cm^{-2}$). It is the external manifestation of an object's thermal state.

Radiant temperature   The concentration of radiant flux from a material. It can be measured by remote sensing instruments known as radiometers.

Specific heat   The heat capacity of 1 g of a material.

Stefan-Boltzmann law   The total radiant exitance coming from a surface is related to its internal temperature (from the kinetic motion of its atoms).

Temperature   A measure of the concentration of the kinetic heat, that is, the thermal state of an object. Temperature can be measured by using some sort of instrument.

Thermal capacity   The ability of a material to store heat. It is a measure of the increase in thermal energy content per degree of temperature rise, specifically the number of calories required to raise the temperature by 1°C (unit: $cal \cdot °C^{-1}$).

Thermal conductivity   A measure of the rate at which heat passes through a material of a specific thickness. It is expressed as the calories delivered in 1 sec across a 1 cm^2 area through a thickness of 1 cm at a temperature gradient of 1°C (unit: $cal \cdot cm^{-1} \cdot sec^{-1} \cdot °C$).

Thermal crossovers   The diurnal curves for all materials intersect shortly after dawn and near sunset. These intersections are called "thermal crossovers," the time when radiant temperatures of these materials are the same.

Thermal inertia   A measure of the resistance of a material to temperature change. Expressed as calories per square centimeter per second square root per degrees Celsius (unit: $cal \cdot cm^{-1} \cdot sec^{-1/2} \cdot {}^{\circ}C^{-1}$).

Thermal-IR images   Images that displays variations in tone or color that represent temperature differences.

Thermal infrared multispectral scanner (TIMS)   An airborne thermal sensor operated for NASA by Jet Propulsion Lab. The sensor uses the across-track scanning technology, and breaks the 8 to 12 μm interval into six bands with spatial resolution of 18 m: (1) 8.2–8.6 μm, (2) 8.6–9.0 μm, (3) 9.0–9.4 μm, (4) 9.4–10.2 μm, (5) 10.2–11.2 μm, and (6) 11.2–12.2 μm. TIMS data have been used extensively in studies of minerals in rocks, soils, and volcanoes.

Urban heat island effect   It is first defined in terms of air temperatures to represent the difference between warmer urban and cooler rural areas. This definition has been adapted to looking at the spatial patterns of land surface temperature in a city by using remote sensing thermal-IR data, which is often termed as surface temperature urban heat island.

Wien displacement law   The law governs the relationship between the peak wavelength and radiant body temperature. The hotter the radiating body, the greater is its radiance over its range of wavelengths, and the shorter is its peak emission wavelength.

Review Questions

1. For an area of your interest, such as geology, agriculture, or urban applications, at what time of day should thermal-IR images be obtained? Explain.

2. Thermal-IR images are very useful in military applications, although we do not measure temperatures from them. Justify this statement.

3. List and explain ways we may use thermal-IR images in the studies of urban landscape. What information is necessary to detect heat leaks for houses?

4. Justify the applications of thermal-IR imagery in forestry and agriculture. In the mid-latitude region, which season of year is preferable to acquire thermal-IR imagery for these applications?

5. Interpreting thermal-IR images and temperature distribution over an area is not straightforward. Identify some circumstances where we just need to look at the patterns of relative temperature differences. Can you show other instances where we have to determine absolute temperature values?

6. List and explain the thermal signatures for some common materials in both daytime and nighttime images: standing water, concrete surface, vegetation, moist soils, and rocks.

7. In the daytime thermal-IR imagery, contrasts between heated buildings and streets and areas in shadow create a scene that resembles an aerial photograph. Explain the advantages and disadvantages of thermal-IR images over aerial photographs.

8. Comparing with reflective bands of satellite imagery, thermal-IR bands tend to have lower spatial resolution. Why do scientists think that thermal-IR imagery is still useful?

References

Cassels, V., J. A. Sobrino, and C. Coll. 1992a. On the use of satellite thermal data for determining evapotranspiration in partially vegetated areas. *International Journal of Remote Sensing*, 13:2669–2682.

Cassels, V., J. A. Sobrino, and C. Coll. 1992b. A physical model for interpreting the land surface temperature obtained by remote sensors over incomplete canopies. *Remote Sensing of Environment*, 39:203–211.

Dash, P., F. -M. Gottsche, F. -S. Olesen, and H. Fischer. 2002. Land surface temperature and emissivity estimation from passive sensor data: theory and practice-current trends. *International Journal of Remote Sensing*, 23:2563–2594.

Jensen, J. R. 2007. *Remote Sensing of the Environment: An Earth Resource Perspective*, 2nd ed. Upper Saddle River, N.J.: Prentice Hall.

Kimes, D. J. 1983. Remote sensing of row crop structure and component temperatures using directional radiometric temperatures and inversion techniques. *Remote Sensing of Environment*, 13:33–55.

Larson, R. C., and W. H. Carnahan. 1997. The influence of surface characteristics on urban radiant temperatures. *Geocarto International*, 12:5–16.

Lillesand, T. M., R. W. Kiefer, and J. W. Chipman. 2008. *Remote Sensing and Image Interpretation*, 6th ed. Hoboken, N.J.: John Wiley & Sons.

Lo, C. P., D. A. Quattrochi, and J. C. Luvall. 1997. Application of high-resolution thermal infrared remote sensing and GIS to assess the urban heat island effect. *International Journal of Remote Sensing*, 18:287–304.

Quattrochi, D. A., and J. C. Luvall. 1997. *High Spatial Resolution Airborne Multispectral Thermal Infrared Data to Support Analysis and Modeling Tasks in EOS IDS Project ATLANTA*. Global Hydrology and Climate Center, NASA Marshall Space Flight Center. http://www.ghcc.msfc.nasa.gov/atlanta/.

Quattrochi, D. A., and M. K. Ridd. 1998. Analysis of vegetation within a semi-arid urban environment using high spatial resolution airborne thermal infrared remote sensing data. *Atmospheric Environment*, 32:19–33.

Sabins, F. F. 1997. *Remote Sensing: Principles and Interpretation*, 3rd ed. New York: W.H. Freeman and Company.

Short, N. M., Sr. 2009. *The Remote Sensing Tutorial*, http://rst.gsfc.nasa.gov/. Accessed January 25, 2011.

Snyder, W. C., Z. Wan, Y. Zhang, and Y.-Z. Feng. 1998. Classification-based emissivity for land surface temperature measurement from space. *International Journal of Remote Sensing*, 19:2753–2774.

Sobrino, J. A., J. C. Jiménez-Muñoz, and L. Paolini. 2004. Land surface temperature retrieval from LANDSAT TM 5, *Remote Sensing of Environment*, 90:434–440.

Weng, Q. 2009. Thermal infrared remote sensing for urban climate and environmental studies: methods, applications, and trends. *ISPRS Journal of Photogrammetry and Remote Sensing*, 64(4):335–344.

Weng, Q., D. Lu, and J. Schubring. 2004. Estimation of land surface temperature-vegetation abundance relationship for urban heat island studies. *Remote Sensing of Environment*, 89(4):467–483.

CHAPTER 10

Active
Remote
Sensing

10.1 Introduction

As we discussed in Chap. 6, both radar and Lidar are active remote sensing techniques. There are a few commonalities in the active remote sensing technology. Both radar and Lidar send out their own electromagnetic energy toward ground targets, which essentially are not affected by atmospheric scattering and most weather conditions. Second, transmitted energy is then interacted with the targets, and is bounced or scattered off the ground surface in different ways. The illumination direction, or viewing geometry, can be modified to enhance detection and identification of the features of interest. Third, the backscattered or reflected signals are received by the receiver, and recorded as remote sensing image data for subsequent processing. Finally, both radar and Lidar measure the roundtrip time of the signal, in addition to signal strength. Therefore, they are considered as ranging sensors.

The differences between radar and Lidar techniques are also apparent. A fundamental difference is that radar uses long-wavelength microwaves, while most Lidar utilizes the wavelength centering at 1064 nm in the near-IR portion of the electromagnetic spectrum. Radar imagery can be collected by using different wavelength and polarization combinations to obtain different and complementary information about the targets. Radar image data have long been employed to map and assess Earth's resources from both airborne and spaceborne sensors. Lidar is a more recently matured remote sensing technique that allows the measurement of the heights of objects with reasonably good accuracy. Most Lidar systems are currently airborne. However, both radar and Lidar technologies have rapidly developed in recent years, making them new frontiers in remote sensing.

10.2   Radar Principles

A typical radar system consists of the following components: a pulse generator, a transmitter, a duplexer, a receiver, an antenna, and an electronics system to process and record the data (Fig. 10.1). The pulse generator discharges timed pulses of microwave energy, which serve two purposes: controlling the bursts of energy from the transmitter and synchronizing the recording of successive energy returns to the antenna (Sabins, 1997). Successive pulses of microwave are focused by a directional antenna into a stream (radar beam). The radar beam illuminates the surface obliquely at a right angle to the flight line of the aircraft or satellite. When a portion of the transmitted energy is reflected or backscattered from ground targets, the same antenna also picks up returned pulses and sends them to the receiver. The receiver amplifies and converts them into video signals and records the timing of the returning pulse, which is used to determine the position of ground features on the image. A duplexer is needed to separate the transmitted and received pulse by blocking reception during transmission and vice versa. A recording device stores the returned pulse for later processing by computers that process the signal into images.

Radar antennas are usually mounted on the underside of the aircraft or spacecraft and direct their beam to the side of the platform in a direction normal to the flight path. For the aircraft, this mode of

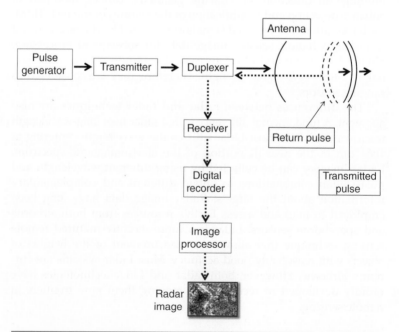

FIGURE 10.1   A side-looking radar system with all components. (*After Sabins, 1997.*)

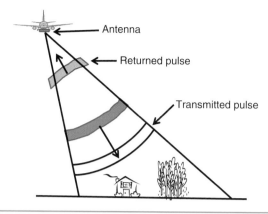

FIGURE 10.2   Operating principle of side-looking airborne radar.

operation is called side looking airborne radar, SLAR (Fig. 10.2). Two basic types of radar systems, real aperture radar and synthetic aperture radar, have been used. Aperture is the opening used to collect the reflected energy, and is thus analogous to the field of view in optical remote sensing or to shutter opening in the case of a camera (Freeman, 1996). A real aperture radar (RAR) system, also known as "brute force" radar, operates with an antenna that has a discrete physical length. This type of radar produces a beam of non-coherent pulses, with narrow angular beam width, and maximizes the antenna length to obtain the desired resolution in the azimuthal (flight line) direction (Sabins, 1997; Short, 2009). For a SLAR radar, a long antenna (typically 5 to 6 m) is used, which is usually shaped as a section of a cylinder wall (Short, 2009).

A major problem with the RAR is its limitation in the system spatial resolution, which relies on the length of the antenna. Increase in the size of the antenna and associated weight often create an operational difficulty in putting it on an aircraft. For a spacecraft, it is not feasible to carry a very long antenna, which is required for imaging the Earth's surface at suitable high resolution. Synthetic aperture radar (SAR) is designed to overcome the limitation in antenna length. A synthetic aperture is constructed by moving a real aperture or antenna through a series of positions along the flight line (Freeman, 1996). As the platform advances, SAR sends its pulses from different positions (Fig. 10.3). At any instant in time, the transmitted beam propagates outward within a fan-shaped plane, perpendicular to the flight line, generating a "broad beam" through which targets remain to be illuminated for a longer period of time (Short, 2009). All echoes from a target for each pulse are recorded during the entire time that the target is within the beam. By integrating the pulse echoes into a composite signal, the movement of the antenna helps to increase the effective length of the antenna, simulating a real aperture (Short,

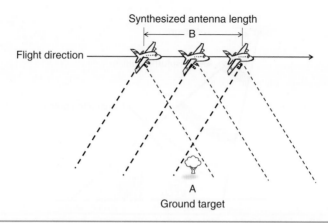

Synthesized antenna length

Flight direction

B

A
Ground target

FIGURE 10.3   Diagram of a synthetic aperture radar. (*After Canada Centre for Remote Sensing, 2007.*)

2009). It is possible through appropriate SAR processing to simulate effective antenna lengths up to 100 m or more. SAR has now become a mature technique that most airborne and spaceborne radars use to generate images with fine details.

The pulses of electromagnetic energy from the transmitter possess specific pulse length and wavelength. Each pulse lasts for only microseconds, typically about 1500 pulses per second. Pulse length is the distance traveled during the pulse generation. As discussed in Chap. 6, radar operates in the microwave region of the electromagnetic spectrum, specifically in the frequency interval from 40,000 to 300 megahertz (MHz). The radar pulse normally covers a small band of frequencies, centered on the frequency selected for the radar system. Table 10.1 lists the common wavelength bands, along with their corresponding frequencies, in pulse transmission. The bands are designed by random letters, which were originally selected to ensure military security during World War II and remain used today for convenience. It should be noted that wavelength may change when

Band	Frequency (MHz)	Wavelength (cm)
Ka	40,000–26,000	0.8–1.1
K	26,500–18,500	1.1–1.7
X	12,500–8000	2.4–3.8
C	8000–4000	3.8–7.5
L	2000–1000	15.0–30.0
P	1000–300	30.0–100.0

TABLE 10.1   Radar Bands and Specifics

passing through media of different materials, but frequency remains constant. Therefore, frequency is a property that is more fundamental in radar remote sensing. Wavelengths are used to designate various radar bands because interpreters can more easily visualize wavelengths than frequencies, and wavelengths are also used to describe the visible and IR spectra (Sabins, 1997). Among the radar bands, Ka, K, and Ku bands are very short wavelengths used in airborne radar systems in the 1960s and early 1970s, but they are uncommon today. X-band is widely used on airborne systems for military reconnaissance and terrain mapping. Other radar bands have also been used in various radar experiments (Canada Centre for Remote Sensing, 2007). For example, C-band is common on many airborne research systems (NASA AIRSAR) and spaceborne systems (Imaging Synthetic Aperture Radar onboard ERS-1 and -2 and RADARSAT). The Russian ALMAZ satellite uses S-band. While L-band has been used with NASA's SEASAT, Japanese JERS-1 satellites, and other NASA airborne systems, P-band, the longest radar wavelength, is used on NASA experimental airborne research system (e.g., AIRSAR/TOPSAR). The shorter bands, such as C band, are more useful for imaging ocean, ice, and some land features. The longer wavelength bands, such as L band, are more penetrating into vegetation canopies and are thus more suitable in forest studies.

Another transmission characteristic of radar signals is polarization. Polarization defines the geometric plane in which the electric field is oscillated (Fig. 10.4). If the electric field vector oscillates along a direction parallel to the horizontal direction of wave propagation, the beam is said to be "horizontally" (H) polarized. In contrast, if the electric field vector oscillates along a direction perpendicular to the horizontal direction of wave propagation, the beam is said to be "vertically" (V) polarized. Similarly, the radar antenna may be designed to receive either horizontally or vertically polarized backscattered energy, or both polarization signals. Most reflected pulses return with

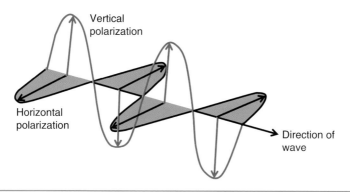

Figure 10.4   Radar polarization.

the same direction of electric field vibration as the transmitted pulse, which is known as like-polarization or parallel-polarization. This is the case for an HH (horizontal transmit and horizontal receive) or VV (vertical transmit and vertical receive) polarization pairing of the transmitted and returned signals. However, upon striking ground targets, a portion of the pulses may be depolarized and returned with different vibration directions. A second antenna may be set to pick up a cross-polarization pulse that is orthogonal to the transmitted direction, leading to either a VH (vertical transmit and horizontal receive) or an HV (horizontal transmit and vertical receive) mode. Vegetation tends to produce a strong depolarization due to multiple reflection of leaves, twigs, and branches (Sabins, 1997), and show different degrees of brightness in HV or VH images. Radar systems that collect images with different polarization and wavelength combinations may provide contrasting and complementary information about the ground targets.

10.3   Radar Imaging Geometry and Spatial Resolution

Knowledge of the geometry of radar data collection is essential in understanding the spatial resolution of radar imagery and in interpreting the geometric characteristics of the imagery. Figure 10.5 shows basic geometry in radar data collection. As the platform advances along the flight direction, a radar beam is transmitted obliquely to illuminate a swath. The swath is offset from nadir, the point directly beneath the platform. The across-track dimension of the swath defines a range that is perpendicular to the flight direction, while the along-track dimension parallel to the flight direction refers to azimuth. The

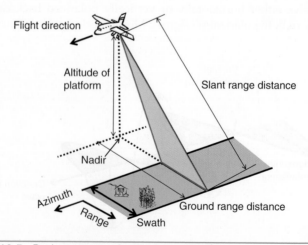

FIGURE 10.5   Basic geometry of radar data collection.

portion of the image swath closest to the nadir is called the near range, whereas the portion of the swath farthest from the nadir refers to the far range. The house in Fig. 10.5 is located closer to the near range than the bush. Radar is a distance-measuring system, and it measures the distance to ground features in slant range rather than in the horizontal distance along the ground. As seen in Fig. 10.5, slant range distance is the radial line of sight distance between the radar and each target on the surface. The horizontal distance along the ground between each measured feature and the nadir defines ground range distance. These two distances along with the altitude of plat-form form a right triangle. This unique geometry is important because we may compute ground range distance based on the slant range dis-tance and platform altitude by using trigonometry, and we may use it to correct slant range distortion in radar images that will be discussed in detail in the following.

In addition, three important angles are also essential to the geo-metrical nomenclature of radar data collection. The angle at which the radar "looks" at the surface refers to look angle (Fig. 10.6). It mea-sures from the nadir to an interested feature. Complementary to the look angle is depression angle, which is defined as the angle between a horizontal plane and a given slant range direction. As seen in Fig. 10.6a, the depression angle becomes smaller when moving out-ward from near to far range. The incident angle more correctly describes the relationship between a radar beam and a surface on which it strikes. The incident angle refers to the angle between the radar beam and a line perpendicular to the surface. It increases as the radar beam moves across the swath from near to far range. Over a flat horizontal surface, the incident angle is approximately equal to the look angle. However, when the terrain is inclined, there is no such relationship between the two angles (Fig. 10.6b).

The spatial resolution of radar imagery is modulated by the specific properties of radar beam and geometrical effects, and is manifested by

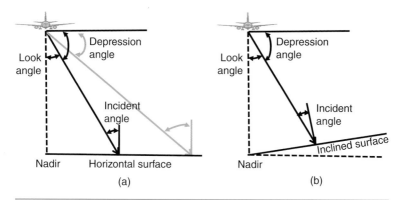

FIGURE 10.6   Incident angle, depression angle, and look angle.

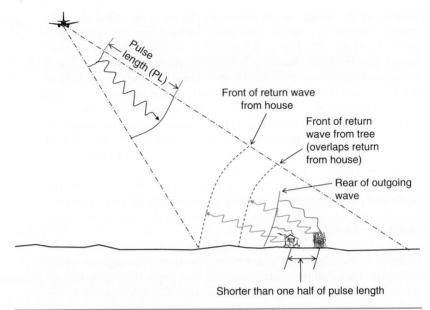

Front of return wave
from house

Front of return
wave from tree
(overlaps return
from house)

Rear of outgoing
wave

Shorter than one half of pulse length

Figure 10.7   Range resolution and pulse length. (*After Lillesand, Kiefer, and Chipman, 2008.*)

the dimensions of the ground resolution cell. The combination of range resolution and azimuth resolution defines the ground-resolution cell size. The range resolution is dependent on the pulse length. Two objects on the surface will be resolved if their separation in the range direction is greater than one-half of the pulse length (Fig. 10.7). Conversely, if the distance between two objects were less than one-half of the pulse length, the radar antenna would receive two overlapped reflected signals, giving rise to blurring in their images. However, when projected onto ground range coordinates, the resolution in ground range would be dependent on the depression angle. For a fixed slant range resolution, the ground range resolution varies inversely with the cosine of the depression angle, implying that a ground range resolution cell gets smaller outward from near to far range. The equation for computing ground range resolution (R_r) is as follows:

$$R_r = \frac{\tau \cdot c}{2\cos\gamma} \tag{10.1}$$

where τ is the pulse length (in microseconds), c is the speed of light (3×10^8 m/sec), and γ is the depression angle. Figure 10.8 shows two targets (A1 and A2) 25 m apart in the near range and another two targets in the far range at the same separation distance (B1 and B2).

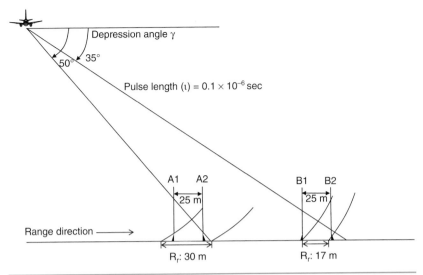

Figure 10.8 Range resolution at two different depression angles. Targets A1 and A2 cannot be resolved in the near range, but targets B1 and B2 are resolvable at the far range. (*After Sabins, 1997.*)

We can compute the ground range resolution at the near range with a depression angle of 60° and a pulse length of 0.1 μsec as:

$$R_r = \frac{(0.1 \times 10^{-6} \, sec) \cdot (3 \times 10^{8} \, m \cdot sec^{-1})}{2 \cos 60°}$$

$$= \frac{30 \text{ m}}{2 \times 0.5}$$

$$= 30 \text{ m}$$

Targets A1 and A2 cannot be separated because their distance (25 m) is smaller than the range resolution distance of 30 m. Targets B1 and B2 are imaged at a depression angle of 35°, and the ground range resolution is computed to be 17.32 m. Therefore, these two targets are resolvable because they have a wider distance than the ground resolution cell.

Both the angular width of radar beam and slant range distance determine the resolution cell size in the azimuth (flight) direction (Fig. 10.9). Two ground objects must have a distance wider than the beam width in the flight direction to be resolved. As illustrated in Fig. 10.9, the radar beam has a fan-shape, which is narrower in the near range than in the far range. Therefore, the azimuth resolution worsens (becomes coarser) outward from near to far range. In the

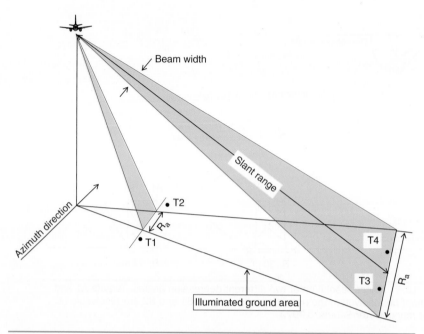

FIGURE 10.9   Azimuth resolution as a function of slant range and antenna beam width. Targets 1 and 2 are resolvable, but targets 3 and 4 are not.

illustration, targets 1 and 2 in the near range are resolvable, but targets 3 and 4 at the far range would not be resolved. The radar beam width is directly proportional to the wavelength of the transmitted energy and is inversely proportional to the antenna length. The equation for computing azimuth resolution (R_a) can be expressed as:

$$R_a = S \cdot \frac{\lambda}{L} \tag{10.2}$$

where S is slant range distance, λ is the wavelength, and L is the antenna length. To improve the azimuth resolution, radar operation needs to be restricted to relatively short range, longer antenna, and the use of short wavelengths. For a given depression angle, the slant range distance is controlled by the height of the aircraft/spacecraft platform ($S = \text{height}/\sin\gamma$). This restriction often implies a low altitude operation for radar systems. Moreover, shorter wavelengths are subject to the impact of weather. As previously discussed, a longer antenna can cause some logistic problems when trying to carry it onboard. Because of these limitations with RAR, SAR has been developed to generate a virtual antenna length so that finer resolution can be achieved.

10.4 Characteristics of Radar Images

The nature of a radar imaging system as a distance measuring device and side-looking illumination have significant effects on the geometrical characteristics of resultant images. The peculiar characteristics we discuss here include slant-range distortion, relief displacement, and parallax. In addition, we will discuss radar image speckle, which is not strictly a geometrical characteristic, but is a noteworthy radar image property.

Radar images can be recorded in either slant-range or ground-range format. In a slant-range image, the distance between the antenna and the target is measured and the spacing between pixels is directly proportional to the time interval between received pulses (Lilesand et al., 2008). In a ground-range image, the distance between the nadir line and the target is measured and the spacing of pixels is directly proportional to their distance on the chosen reference plane. In the slant-range image, the scale varies from near to far range. Targets in the near range appear compressed, while those in the far range appear expanded. Figure 10.10 illustrates the relationship between the slant-range and ground-range image format. Target A1, in the near range, and target B1, in the far range, have the same dimensions on the ground, but their apparent dimensions in the slant range are considerably different. Target A2 is much smaller than target B2.

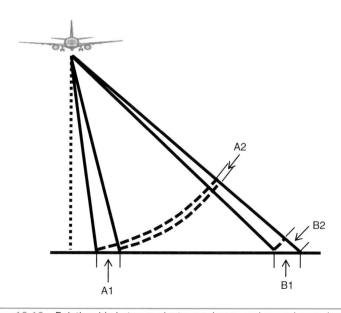

FIGURE 10.10 Relationship between slant-range image and ground-range image.

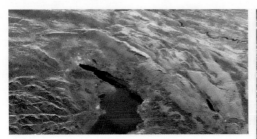

Figure 10.11 A SEASAT image scene of Geneva, Switzerland. The slant range image is displayed on the left, while the corrected ground range image is on the right. (*Image courtesy of European Space Agency.*)

For a fixed swath width, this image scale distortion reduces with an increase in the height of the platform (Lillesand et al., 2008). Therefore, radar images from a spacecraft possess substantially smaller scale distortion than those from aircrafts. However, it is possible to correct the slant-range scale distortion by using trigonometry. Ground-range distance can be calculated from the slant-range distance and the platform altitude, and slant-range images can be converted to the proper ground-range format. Figure 10.11 shows an example of a SEASAT image of Geneva, Switzerland. The slant-range image is displayed on the left, while the corrected ground-range image is displayed on the right.

When radar pulses hit vertical features, such as mountains and tall buildings, the resultant radar images are subject to geometric distortions due to relief displacement. Similar to line scanner imagery, this displacement is one-dimensional and perpendicular to the flight path, i.e., across-track distortion (Lillesand et al., 2008). The unique characteristic of radar imagery is that this relief displacement is reversed with targets being displaced toward, instead of away from the sensor. Radar layover and foreshortening are two forms of relief displacement.

The time delay between the radar echoes received from two different points determines their relative distance in an image. When the radar beam reaches a tall vertical feature, the top of the feature is often reached before its base. The return signal from the top of the feature will be received before the signal from the bottom. As a result, the top of the feature is displaced toward the nadir from its true position on the ground, and appears to "lay over" the base of the feature (Fig. 10.12).

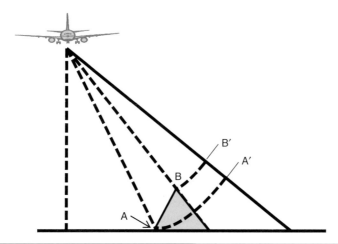

Figure 10.12 Radar layover. The top of the feature is displaced toward the nadir from its true ground position, and appears to "lay over" the base of the feature (B' to A').

The layover zones, facing radar illumination, appear to be bright features on the image due to the low incidence angle. When a vertical feature is flatter, layover will not occur. Instead, the radar pulse will reach the base of a feature first and then the top. Foreshortening will occur, as illustrated in Fig. 10.13. The slope (A to B) will appear compressed and the length of the slope will be represented incorrectly

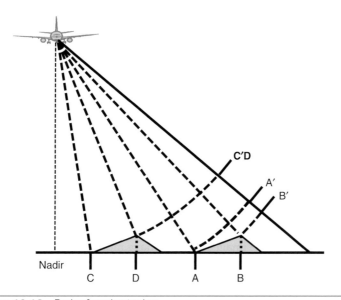

Figure 10.13 Radar foreshortening.

(A' to B'). The angle of the hillside or mountain slope in relation to the incidence angle of the radar beam determines the severity of foreshortening (Canada Centre for Remote Sensing, 2007). When the radar beam is perpendicular to the slope such that the slope, the base, and the top of a feature are imaged simultaneously, the maximum degree of foreshortening occurs, in which the length of the slope will be reduced to an effective length of zero in the slant range (Canada Centre for Remote Sensing, 2007). Moreover, larger depression angles foreshorten more than smaller angles. A larger depression angle makes the slope length appear shorter, and thus increases the degree of foreshortening. Radar images acquired from aircrafts are more likely to show layover and foreshortening than those from spacecrafts are. Because, at higher altitudes, satellites have higher depression angles, the radar beam reaches fore and back slopes at approximately the same time, minimizing the differences between near- and far-range timing and imaging (Short, 2009). Both relief displacements are more pronounced in the near range of the image swath than in the far range because the depression angle is larger in the near range.

Both foreshortening and layover result in radar shadow. It occurs when the radar beam is not able to illuminate the ground surface, such as mountain slopes facing away from the radar beam. The shadowing effect becomes less severe as the incidence angle increases from near to far range and the radar beam looks more and more obliquely at the surface. Apparently, there is an inverse relationship between the relief displacement and the shadowing. Radar images acquired with larger incident angles have more pronounced foreshortening and layover but less shadowing, and vice versa.

When two radar images are taken for the same area, different relief displacements create image parallax on radar images. Similar to the concept of stereo mapping using aerial photography (Chap. 5, Sec. 5.6), these radar images can be utilized to view terrain stereoscopically and to measure the heights of features. Radar images can be taken through multiple passes from the same side (i.e., same-side stereo), with the second time moving over nearly the same ground surface as the previous pass (Fig. 10.14a). It is also possible to acquire a stereo pair of images at two incidence angles by taking different flying heights on the same flight line. Alternatively, stereo radar images can be acquired from the flight lines on the opposite sides of a terrain strip (opposite-side stereo; Fig. 10.14b). The same-side stereo is preferred over the opposite-side stereo because the former is consistent in terms of the direction of illumination and displacement between the two images. The opposite-side configuration may yield significant contrast in the images and may have more shadowing problem. The science of obtaining reliable measurements from stereo radar images has been termed radargrammetry, as an analog of photogrammetry for stereo aerial photographs. In addition, a stereo pairing is

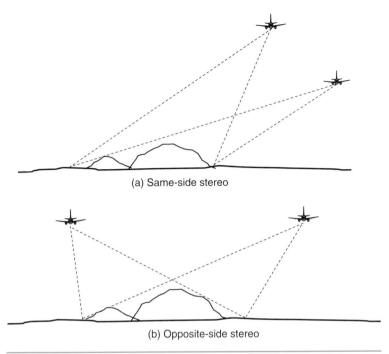

(a) Same-side stereo

(b) Opposite-side stereo

FIGURE 10.14   Stereoscopic acquisitions of radar images.

achievable during a single pass by having two separate but coordinated radar transmitter/receiver systems operating simultaneously from their moving platform (Short, 2009). The radars can be of the same band or different bands. The returned signals will show phase differences, meaning that two waves with the exact same wavelength and frequency travel along in space but with slightly different starting points. This phenomenon is called radar interference, as illustrated in Fig. 10.15. The returned signals are processed by using the technique known as interferometry, which works well with the signals of coherent radiation. The phase difference between adjacent ground cells represents the variations in height. The information can be used to derive topographic information and produce 3D imagery of terrain height. Figure 10.16 shows an SRTM image of Madrid, Spain. Elevation data used in this image were acquired with the technique of interferometry. SRTM used SIR-C/X-SAR, and was designed to collect 3D measurements of the Earth's surface.

Even when radar interference is not purposely engineered for acquiring images, random interference may occur from multiple scattering returns within a single ground resolution cell. Radar pulses traveling from the antenna to the surface and back may have slightly different distances. This difference can result in small phase shifts in

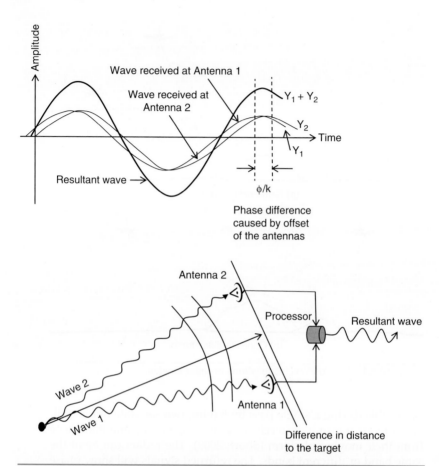

Figure 10.15   Concept of radar interference. (*After Sabins, 1997.*) (*a*) Origin of interference from two received radar waves that are not in phase; (*b*) radar interferometer system. One antenna transmits radar pulses, but they are received by two antennas placed at different distances from the target, causing phase difference. Two return waves combine by a processor to generate interference.

the returned radar signals. The returning signals can either enhance each other, known as constructive interference, or cancel each other out, referred to as destructive interference (Lillesand et al., 2008). These forms of radar interference give rise to a seemingly random pattern of brighter and darker pixels in the resultant images, leading to a grainy "salt and pepper" texture in an image known as radar speckles. Radar speckles are, in fact, not so "random" because they are directly related to ground geometrical irregularities. For example, plowed fields can cause radar signal backscatter in various directions, as can individual blades of grass within each ground resolution cell. Both types of ground surfaces may generate radar speckles. Figure 10.17 shows a SAR image of some agricultural fields located

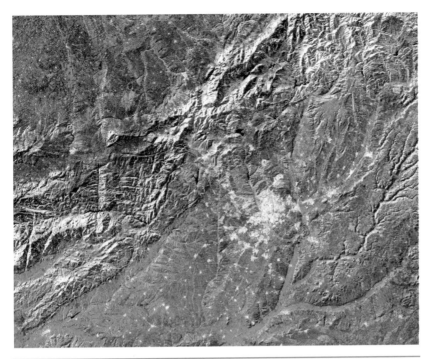

Figure 10.16 Shuttle Radar Topography Mission (SRTM) image of Madrid, Spain. Situated in the middle of the Meseta and at an elevation of 646 m (2119 ft) above sea level, Madrid is the highest capital city in Europe. Elevation data used in this image were acquired by the SRTM aboard the Space Shuttle Endeavour, launched on February 11, 2000. SRTM used Spaceborne Imaging Radar-C/X-Band Synthetic Aperture Radar (SIR-C/X-SAR), and was designed to collect 3D measurements of the Earth's surface. (*Courtesy of NASA/JPL/Caltech.*)

along the Tiber River north to Rome, in the central part of Italy. The homogeneous patches representing the fields have high variability in backscattering due to speckles.

Speckle degrades the quality of a radar image and may make interpretation more challenging. Speckle can be reduced with either the technique of spatial filtering or multi-look processing. Spatial filtering consists of moving a window of a few pixels in dimension (e.g., 3 × 3 pixels) over each pixel in the image, applying a mathematical operation to the pixel values within that moving window, and creating a new image where the value of the central pixel in the window is the result of the mathematical operation. An example is the mean filter. More often, spatial filtering cannot completely remove speckles. Multi-look processing is thus a preferable technique. It divides the radar beam into several portions, and each portion provides an independent "look" at the imaged scene. Each of these "looks" produces an independent image, which may be subject to speckle. By averaging

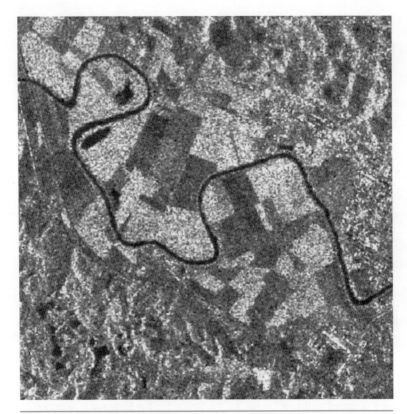

Figure 10.17 An example of radar speckle. The SAR image was acquired on April 21, 1994 over Tiber Valley, showing agricultural fields located along the Tiber River north to Rome, in the central part of Italy. The homogeneous patches representing the fields have high variability in backscattering due to the speckle noise. (*Courtesy of European Space Agency.*)

all images, a smooth image can be produced. Multi-looking is usually done during data acquisition, while spatial filtering is done after image acquisition. In both cases, speckle is reduced at the expense of the spatial resolution of radar imagery.

10.5   Radar Image Interpretation

The pixel values of radar images manifest the radar backscatter for that area on the ground. Bright features mean that a large fraction of the transmitted energy was reflected back to the radar, while dark features indicate that little energy was reflected. The intensity of radar backscatter varies with a variety of parameters, relating to the particular characteristics of the radar system, such as wavelength, polarization, viewing geometry, etc., as well as to the characteristics of the surface, such as dielectric properties, water content, surface roughness, and feature orientation in respect to radar viewing. Many of these parameters are correlated to each other, meaning that change in one parameter may affect the response of other parameters and thus the intensity of backscatter. We discuss these parameters mainly from the perspective of electrical and geometrical characteristics.

A measure for the electrical characteristics of the Earth's surface materials is the dielectric constant. It is a measure of both the conductivity and reflectivity of the materials, and describes a material's capability to hold a charge and to polarize when subjected to an electric field (Short, 2009). Therefore, the dielectric constant modulates the interaction between radar pulses and the surface materials. At the radar wavelengths, the majority of natural materials yield a dielectric constant of 3 to 8 in their dry condition, but contaminated water may have a value as high as 80. The presence of moisture in soil or vegetation enhances the dielectric constant. For instance, moist soils have dielectric constants typically ranging from 30 to 60. The materials or surfaces with high dielectric constants reflect radar energy more efficiently than those with low constants do. However, radar pulses can penetrate farther into the materials with low dielectric constants. Moist plants and soils reflect radar energy well. Wetter materials or surfaces will appear bright in radar images, while drier targets will be dark. The variations in the intensity of radar returns frequently indicate the differences in soil moisture, all other factors being constant (Lillesand et al., 2008).

Surface roughness refers to subtle irregularities in the Earth's surface that are caused by textural features comparable in size to the radar wavelength, such as leaves and twigs, closely spaced vegetation with various shapes, grass blades, pitted materials, granular soils, gravel, etc. and whose surfaces have dimensional variability on the order of millimeters to centimeters. It is completely different from topographic relief that varies in meters or hundreds of meters (Sabins, 1997). Topographic relief features, such as mountains and valleys, influence radar returns in another way, and may generate bright patches and shadows in radar images (refer to Sec. 10.4). Surface roughness determines how radar pulses interact with the surfaces or targets, and thus plays an important role in determining the overall brightness of a radar image.

Whether a surface appears rough or smooth to a radar beam depends on the wavelength and the depression angle of the antenna. The Rayleigh criterion considers a surface smooth if the following relationship is satisfied:

$$h < \frac{\lambda}{8\sin\gamma} \tag{10.3}$$

where h is the height of surface irregularities or roughness, λ is radar wavelength, and γ is depression angle. A smooth surface, acting as a specular reflector, will reflect all or the majority of the incident energy away from the radar and results in a weak return. In contrast, if the height variation of a surface exceeds the wavelength of sensing divided by eight times the sine of the depression angle, that surface is considered as "rough" and to act as a diffuse reflector. Radar energy will be scattered equally in all directions, and the surface will return a significant portion of the energy to the radar. A surface of intermediate roughness reflects specularly a portion of the energy and diffusely scatters the remaining. From Eq. (10.3), it is apparent that a surface may appear smooth on L band radar, but be rough on X band radar (Fig. 10.18). For a given surface and wavelength, the surface will appear smoother as the depression angle decreases (Fig. 10.19). As we move outward from near to far range, less energy would be returned to the radar and the image would become increasingly darker in tone.

Radar backscatter is also highly sensitive to the relationship between radar viewing and feature orientation, surface slope, and shape. Changes in viewing geometry affect radar-target interactions in different ways, and may result in varying degrees of foreshortening, layover, and shadow, as discussed in Sec. 10.4. Local incident angle manifests this relationship well. Radar backscatter decreases with increasing incidence angles. For low incidence angles,

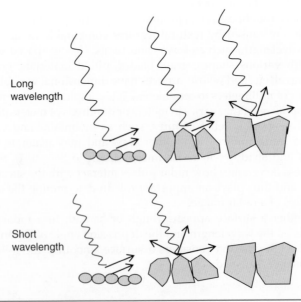

FIGURE 10.18 Dependence of surface roughness on radar wavelength. (*After Arnold, 2004.*)

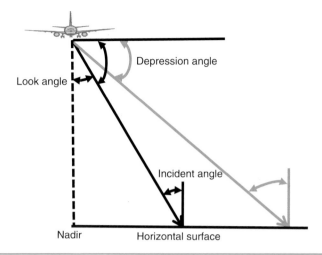

FIGURE 10.19   Effect of depression angle on surface roughness.

such as ranging from 0° to 30°, high backscatter is expected that is dominated by topographic slope; when this angle increases to the range of 30° to 70°, surface roughness would prevail (Lillesand et al., 2008). For angles larger than 70°, shadows would dominate the radar image (Lillesand et al., 2008).

In interpreting radar images, a general rule is that the brighter the image, the rougher the surface being imaged. Flat surfaces, such as smooth water surfaces, airport runways, roads, and freeways, usually appear dark in radar images, reflecting little or no energy back to the radar (referring to specular reflection). Urban areas typically provide strong radar returns due to corner reflection, and thus appear bright in radar images. High-rise buildings in urban areas may line up in such a way that the incoming radar pulses are able to bounce off the streets and then bounce again off the buildings, giving rise to a double reflection that yields a strong return (Freeman, 1996). Various forests are usually moderately rough on the scale of most radar wavelengths, and appear as grey or light grey in a radar image, referring to diffuse reflection (Freeman, 1996). Figure 10.20 illustrate the differences among specular, diffuse, and corner reflection and their signatures in radar images. When the top layer of trees (forest canopy) is very dry and the surface appears smooth to the radar, the radar energy may be able to penetrate below the layer. In that case, radar backscatter may come from the canopy, the leaves and branches further below, and the tree trunks and soil at the ground level. This type of scattering is called volume scattering, which typically consists of multiple bounces and reflections from different components within the volume or medium. Volume scattering may enhance or decrease image brightness, depending on how much of the energy is scattered

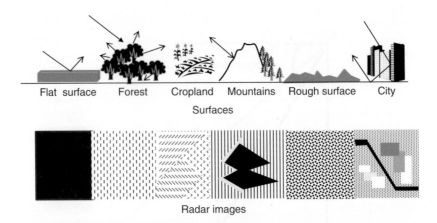

Surfaces

Radar images

FIGURE 10.20   Specular reflection, diffuse reflection, and corner reflection. (*After Freeman, 1996.*)

out of the volume and back to the radar (Canada Centre for Remote Sensing, 2007). Radar wavelengths of 2 to 6 cm are best suited for sensing crops, such as corn, soybeans, and wheat, because the volume scattering predominantly comes from their canopies and the influence of underlying soil is minimal (Lillesand et al., 2008). Longer wavelengths are more appropriate for sensing tree trunks and stems. In addition, more backscatter is returned when the azimuthal direction is parallel with the rows of crops. In fact, the radar look direction has a similar effect on other linear features, such as urban buildings, fences, mountain ranges, ocean waves, and fault lines. If the look direction is perpendicular to the orientation of the feature, then a large portion of the incident energy will be reflected back to the sensor and the feature will appear bright. However, if the look direction is more oblique in relation to the feature orientation, then less energy will be returned and the feature will appear darker (Canada Centre for Remote Sensing, 2007).

10.6   Lidar Principles

Lidar is an active sensing system that transmits laser light as a series of pulses (100 per second) to ground targets, from which some of the light is reflected and the time for the pulse to return is measured. Because the laser pulse travels at the speed of light (a known constant), the time can be converted to distance using the following equation:

$$D = C \times \frac{t}{2}$$ (10.4)

where D is the distance from the sensor to the target, C is the speed of light (3×10^8 m/sec), and t is the elapsed time from sending the pulse to detection of the pulse return. The result is then divided by two to get the distance between the sensor and the target because the light must travel to the target and then double the trip back to the sensor. This distance is often referred to as the range. Because Lidar is an active sensing system, data can be acquired both daytime and nighttime as long as the atmosphere is clear. In this sense, Lidar is similar to radar. However, there are some fundamental differences between the two types of sensing systems. Lidar is an optical remote sensing technology, which uses ultraviolet, visible, or near-IR light to image objects, instead of microwave energy. Most airborne topographic mapping lasers utilize the wavelength centering at 1064 nm in the near-IR portion of the electromagnetic spectrum. Moreover, laser light is different from white or ordinary light because laser light is coherent, meaning that its waves are in phase with one another and have the same wavelength (Popescu, 2011).

Lidar systems may be used as a ranging device to determine altitudes of terrain or depths of water (bathymetric Lidar). Furthermore, Lidar sensing technology is also applicable for studying the atmosphere as a particle analyzer (atmospheric Lidar). For topographic mapping applications, Lidar-derived elevation points must be distributed over a large swath on either side of the flight light on the ground. To achieve this goal, Lidar pulses are often directed from the plane using an oscillating mirror. This method of scanning often gives rise to a resultant scan line in a zip-zap pattern and the overall pattern of a dense set of data points arranged in a swath buffering the flight line (see Fig. 6.8, Chap. 6). In addition to a laser scanner, two other technologies are essential to a Lidar system. Differential kinematic GPS techniques are used to determine the precise x, y, and z positions of the platform, while an inertial navigation system (INS) is used to determine the precise orientation of the aircraft and the orientation of the scanning mirror.

The central component of a Lidar system is a laser scanner. It includes a laser source, a laser detector, a scanning mechanism, electronics for timing the pulses and returns, and a computer to process and record the data in real time. The beam of laser light, when it reaches the targets, forms a cone rather than a line (Fig. 10.21). The distribution of beam intensity across the footprint is Gaussian, meaning that the intensity gradually reduces toward the edges of the footprint according to the Gaussian distribution in statistics (bell-shaped distribution). Thus, the width of the beam contains approximately 87% of the energy of the beam (Hecht, 1994). The diameter of the beam (the diameter of footprint) may be computed as follows:

$$D = H \times \gamma \qquad (10.5)$$

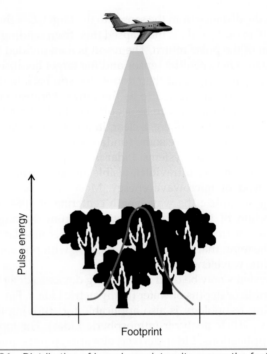

FIGURE 10.21   Distribution of laser beam intensity across the footprint as a Gaussian function.

where D is the laser footprint width (m), H is the altitude above ground level (m), and γ is the beam divergence angle in radians (1 radian = 57.3°). Typical Lidar sensors have a divergence angle between 0.25 to 4 MRAD. Therefore, when imaging from the altitude of 1000 m, the diameter of a footprint for these Lidar sensors will be between 0.25 to 4 m (9.8 in. to 13 ft).

10.7   Characteristics of Lidar Data

It is likely that the same pulse of laser beam strikes several features within the footprint. Each portion of the laser beam may strike features at various heights, and is reflected back at different times, which results in a multi-return Lidar. For example, a laser beam may hit the tree canopy first and result in a strong return, and then a portion of the energy continues to travel downward and hits branches and leaves, generating a second return of the same pulse. When the laser pulse eventually reaches the ground, it may generate a final return. Early Lidar systems can record only one discrete return, either the first peak or the final peak of the reflected energy. Today, commercial Lidar systems are capable of measuring multiple returns, up to five returns per pulse. A filter needs to be applied to detect peaks in the

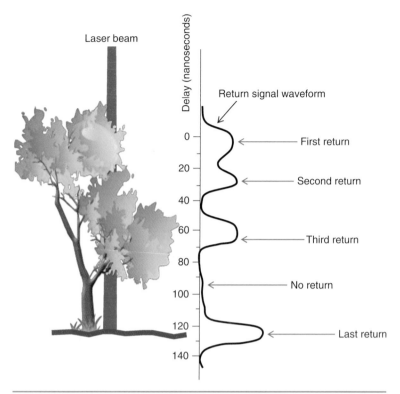

FIGURE **10.22** Waveform and multiple Lidar returns can be generated from a single emitted pulse. Depending on the detection capabilities of the Lidar sensor, up to five discrete returns may be recorded as data points.

reflected waveform and to record the timing of those peaks as discrete "returns" (Fig. 10.22). These discrete Lidar returns can be analyzed and classified to produce information about objects above the ground and the bare ground surface. Among all the returns, the first and last returns are the most important ones in topographic mapping. Figure 10.23 illustrates the first and final Lidar return for a portion of downtown Indianapolis, Indiana, U.S.A. The first return usually provides information about the surface of all ground objects (e.g., treetops, building tops, etc.), and can be used to generate a digital surface model (DSM). In comparison, the last return points hit through the openings among tree branches and leaves to the non-penetrable objects such as the ground and buildings. The last return may be used to generate a digital elevation model (DEM) for the "bare-Earth" surface after removing buildings and vegetation not penetrated by laser pulses.

The entire waveform from the laser pulse can also be recorded, as opposed to discrete returns (Fig. 10.22). Most Lidar systems that

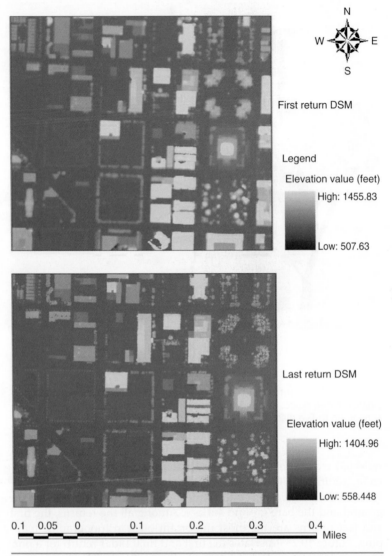

First return DSM

Legend

Elevation value (feet)

High: 1455.83

Low: 507.63

Last return DSM

Elevation value (feet)

High: 1404.96

Low: 558.448

0.1 0.05 0   0.1   0.2   0.3   0.4
                                          Miles

FIGURE 10.23   First and final Lidar return for a portion of downtown
Indianapolis, Indiana, U.S.A. (*Data provided by Indianapolis Mapping and
Geographic Infrastructure System.*)

receive multiple discrete returns need to reset their detectors to prepare for the next pulse to be returned from the same pulse. The time
separation between the two returns, referred to as the reset time or
nominal dead time, varies with sensors, typically ranging from 8 to
10 nsec (Popescu, 2011). This reset time translates into a minimum
range separation of 1.2 to 1.5 m between the recorded returns of the

same pulse (Popescu, 2011). The reset time and the minimum range separation between multiple returns imply that when the ground is covered by tall grass or shrubs with heights less than 1.2 to 1.5 m, the laser beam may provide a return from the top of the canopy but a ground return may not be detectable because no return is recorded within the reset time (Popescu, 2011). The full waveform Lidar can overcome this limitation in data recording. This technique provides more detailed information on vertical structure and more accurate detection of the true ground topography, and thus is very useful in the analysis of vegetation density, mapping live/dead vegetation, forest fuels analysis, and wildlife habitat mapping. In addition, waveform Lidar is not subject to signal processing errors. However, the vast amount of data from the waveform Lidar, approximately 30 to 60 times that from the discrete-returns Lidar, can post a big challenge in data processing. Currently, the waveform Lidar has lower spatial resolution than discrete Lidar.

Similar to radar, most Lidar systems record the intensity of a return pulse, in addition to the time of the return. Lidar intensity reflects largely the maximum strength of the return instead of individual returns or their combination (Baltsavias, 1999). Buildings show a higher intensity than does an asphalt roadway (Fig. 10.24). However, bright roofs yield a stronger return than dark roofs (Fig. 10.24). The

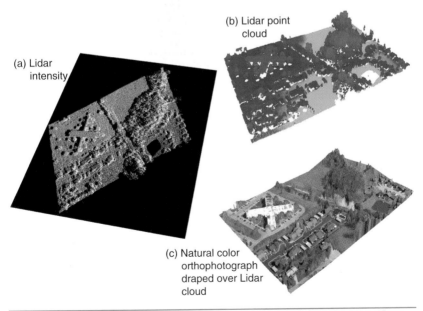

(a) Lidar intensity

(b) Lidar point cloud

(c) Natural color orthophotograph draped over Lidar cloud

FIGURE 10.24   Lidar intensity map provides good clues for ground features. (*Data provided by Indianapolis Mapping and Geographic Infrastructure System.*)

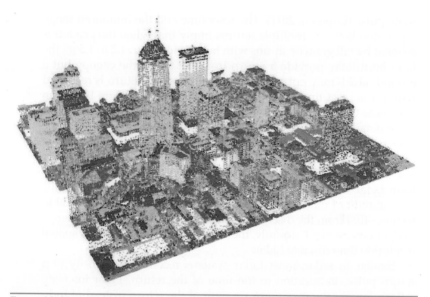

FIGURE 10.25   Lidar point cloud over the downtown area of Indianapolis, Indiana, U.S.A. (symbolized by intensity values of returns). Date density: 1 point/m²; vertical accuracy: 0.15 m; horizontal accuracy: 0.46 m. (*Data provided by Indianapolis Mapping and Geographic Infrastructure System.*)

intensity of return from trees depends on how far the laser pulse can penetrate downward (Jensen, 2007), and thus is a manifest of the result of the interaction between the laser pulse and leaves, twigs, branches, and the trunk. The intensity values can be used to create and to display an image by scaling the minimum and maximum values to an 8-bit (0–255) grayscale palette, superimposing a grid on the Lidar points, and assigning average intensity values to each cell in the grid. Figure 10.25 shows an example of a Lidar data cloud, which covers the downtown area of Indianapolis, Indiana, U.S.A. The data points are scaled in order to display the buildings and other structures.

10.8   Lidar Applications

10.8.1   Topographic Mapping Applications

Since the inception of Lidar systems, Lidar technology has been used widely in many geospatial applications owing to its high-resolution topographic data, short acquisition time, and reasonable cost compared with traditional methods. Unlike other remotely sensed data, Lidar data focus solely on geometry rather than radiometry. Lidar data can provide very good vertical and horizontal accuracies as high as 0.3 m (Webster, Forbes, Dickie, and Shreenan, 2004). Elevation measurements from buildings, trees, or other structures, and bare-Earth are all available from raw Lidar data. Some typical products derivable from Lidar data include DEM, DSM, triangulated irregular network (TIN), and contours (lines of equal elevation). These data with vertical accuracies of 15 to 100 cm (Hill et al., 2000) have become a crucial source for a wide variety of applications related to topographic mapping, such as transportation, urban planning, telecommunications, floodplain management, mining reclamations, landside analysis, and geological studies. These projects may need accurate data of surface elevation, the height of aboveground features (e.g., vegetation, buildings, bridges), or both.

Figure 10.26 shows a DSM and DEM created using Lidar data for the city of Indianapolis, Indiana, U.S.A. By creating 3D images or by using animation techniques, Lidar offers a new potential for visualization of urban structures, trees, river valleys, urban corridors, and many more. Figure 10.27 provides an example of a 3D image for a stadium and nearby highways in Indianapolis. By comparing with the aerial photograph of the same area, it is clear that the Lidar used for the application can reveal subtle differences in elevation and delineate urban structures effectively.

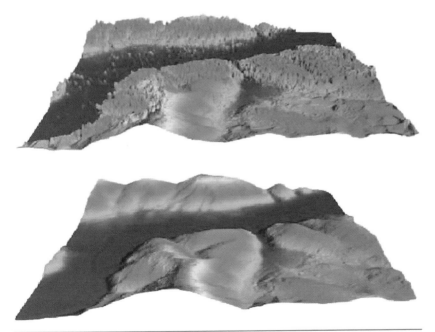

FIGURE 10.26   Digital surface model and digital elevation model generated by using Lidar data for Indianapolis, Indiana, U.S.A. (*Data provided by Indianapolis Mapping and Geographic Infrastructure System.*)

FIGURE 10.27   Lidar data used for revealing subtle differences in elevation and delineating urban structures. (*Data provided by Indianapolis Mapping and Geographic Infrastructure System.*)

Contours may be used for planning and management of drainage, site design for construction of water and sewer pipes, roads, and buildings; and establishing property values. Most contours are generated from aerial photographs and photogrammetry during the process of producing 1:24,000-scale USGS map series. These contours are used later to convert to DEMs. These contours and DEMs often contain significant errors, and therefore it is not feasible to use these products to obtain accurate elevation information. In urban areas, contours may be incomplete in many areas (e.g., discontinuous at buildings, walls, or

FIGURE 10.28   Lidar-derived contours (2-ft interval) overlaid with aerial photograph in a residential area of Indianapolis, Indiana, U.S.A. (*Data provided by Indianapolis Mapping and Geographic Infrastructure System.*)

other features) and out of date (not reflecting new developments), which are not suited for direct measurement and modeling of surface elevation. When covered with dense forestry, elevation values from vegetation and structures may be incorrectly registered to the ground. Lidar technology can provide cost-effective elevation data suitable for engineering use and generating quality contours and DEMs. Figure 10.28 shows Lidar-derived contours (2-ft interval) overlaid with an aerial photograph in a residential area of Indianapolis. As we discussed in Chap. 5, quality DEMs are critical in rectifying orthophotographs. Figure 10.29 shows two different panels of orthophoto covering the same locality. Figure 10.29a is produced using USGS DEM for rectification, in which the highway has not been made planimetric. The orthophoto rectified by Lidar-derived DEM is displayed Fig. 10.29b. The highway and bridge have been made planimetric; in order words, these features have now registered to "right" elevation. In addition, accurate elevation data is crucial to predict flood risks, especially in areas characterized by little terrain relief. The Federal Emergency Management Agency's (FEMA) mapping requirements now allow for use of Lidar data to create new flood insurance studies and associated flood insurance rate maps (Hill et al., 2000).

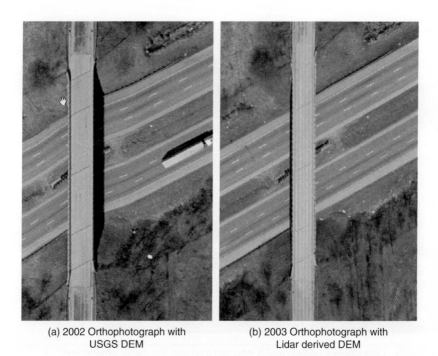

(a) 2002 Orthophotograph with
USGS DEM

(b) 2003 Orthophotograph with
Lidar derived DEM

FIGURE 10.29   A comparison of orthophotos produced using USGS DEM for
rectification (*a*) with the orthophoto rectified by Lidar-derived DEM (*b*). In panel (*b*),
the highway and bridge have been made planimetric, while in panel (*a*), apparent
displacement occurs to the highway and nearby features. (*Data provided by
Indianapolis Mapping and Geographic Infrastructure System.*)

Feature extraction (e.g., buildings and roads) based on Lidar
data is extensive. Lidar data show great potential in the extraction
of buildings and other features in urban areas in comparison with
photogrammetric techniques because elevation data can be derived
quickly and at high resolution (Miliaresis and Kokkas, 2007). The
two images in Fig. 10.30 are computerized visualizations of elevation information of the World Trade Center in July 2000 and on September 15, 2001 after the September 11 attack. These maps were
produced using an airborne Lidar system that provides a vertical
accuracy of 6 in.

10.8.2   Forestry Applications

Researchers working on monitoring and predicting forest biomass
characteristics regard Lidar as a critical data source (Popescu et al.,
2003). Lidar can be used to measure the 3D characteristics of plant
canopies and to examine the vertical structure of vegetation communities. Since different species of trees reflect Lidar pulses differently, identification of dominant species based on crown shapes
may also be possible. In fact, Lidar data have been used to measure

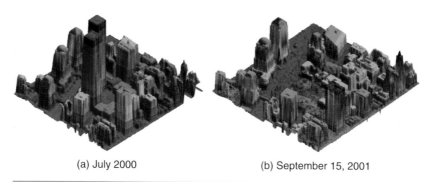

(a) July 2000   (b) September 15, 2001

FIGURE 10.30   Lidar image of the World Trade Center and nearby buildings before the September 11, 2001 attack and on September 15, 2001 after the attack. (*Courtesy of EarthData International.*)

directly such forest parameters as stand density, tree height, crown width, and crown length. By using modeling techniques such as regression, Lidar data have also been employed to estimate forest volume, biomass (the living or dead weight of organic matter in a forest expressed in units such as living or dead weight, wet or dry weight, ash free-weight, etc.), diameter at breast height (DBH), and basal area (the cross-sectional area of all the stems of a species or all the stems in a stand measured at breast height and expressed per unit of land area). These parameters are crucial for understanding the composition, structure, and patterns of a vegetation community, thus contributing to our knowledge of forest conditions and stress, wild fire risks and wildlife habitats, and resource inventory and planning.

Most forestry inventories need information about the composition of tree species, stand size and height, and crown density to calculate volume and site potential and to assess the need for silvicultural treatments. Conventional methods for inventories by aerial photo interpretation, fieldwork, and extrapolation of information through sampling of forest plots are labor intensive and expensive. Passive satellite sensors have a critical limitation in quantifying vegetation characteristics because they cannot measure the understory and vertical characteristics of a forest. Lidar techniques allow 3D analysis of forest structure components such as crown density, crown volume, stand height, and tree density over large areas (Lefsky et al., 1999; Naesset, 2002; Popescu and Wynne, 2004). In Fig. 10.31, Lidar data, acquired in the spring of 1999, are used to map the Capitol State Forest, near Olympia, Washington, U.S.A. This type of survey is useful for investigating forest conditions and resource planning. In another example of Lidar application, Lidar point cloud data show an excellent potential to identify tree shape and height. In Fig. 10.32a, 6-in. resolution of natural color orthophoto taken in March 2010 is displayed to show Indy Military Park, Indianapolis, Indiana, U.S.A. The photos were taken in the leaf-off season, so trees with dark shadows

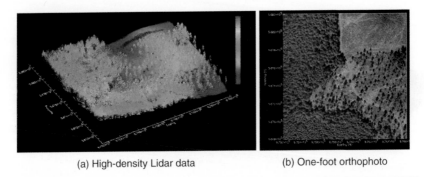

(a) High-density Lidar data   (b) One-foot orthophoto

FIGURE 10.31   Lidar image of Capitol State Forest, near Olympia, Washington, U.S.A. Data acquired in the spring of 1999 covering 5.2 km². (*Source: http://www. westernforestry.org/wmens/m2002/andersen.ppt.*)

are conifers and those without clear shadows are deciduous (Sycamores). In Fig. 10.32*b*, various elevations from low to high are shaded with colors of blue, light blue, green, yellow, and red in the Lidar point cloud (1-m post spacing) map. The Lidar data were acquired in the same season, December 2009. By comparing the two panels of Fig. 10.32, we can clearly separate conifer from deciduous trees.

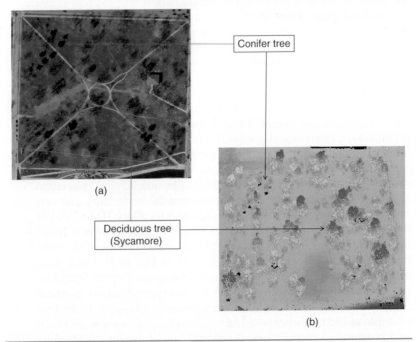

FIGURE 10.32   Lidar data used to identify tree shape, height, and life form. (*Lidar data provided by Jim Stout, Indianapolis Mapping and Geographic Infrastructure System, Indiana, U.S.A.*)

10.9   Summary

In this chapter, we discuss two active remote sensing techniques—
radar and Lidar—widely used now in imaging Earth's resources and
landforms. Each technique was treated in detail with the data collection method and resultant image characteristics, which are followed
by the principles of data interpretation and analysis with practical
examples. The intensity of radar backscatter, manifesting the interaction between transmitted energy and ground targets, is controlled by
transmission characteristics of radar signals, including wavelength,
polarization, and viewing geometry, and the characteristics of surface
materials and structures, which may be composed of dielectric properties, water content, surface roughness, and feature orientation in
respect to radar viewing. The geometry of radar imagery is different
from that of vertical photography and electro-optical sensors. As a
distance-measuring device with side-looking illumination, radar
yields various effects on resultant images, such as slant-range distortion, relief displacement, and parallax. The principle of parallax
can be utilized to produce stereo radar images and thus generate
topographic maps, similar to stereo mapping in photogrammetry.
Topographic maps can also be generated based on the technique of
radar interferometry, which creates a stereo pairing during a single
pass. In interpreting radar images, a general rule is that the rougher
the surface being imaged, the brighter the image. Flat surfaces usually appear dark in radar images due to specular reflection. Urban
areas typically provide strong radar returns due to corner reflection,
and thus appear quite bright in radar images. Various forests are usually moderately rough and appear as grey or light grey in a radar
image because of diffuse reflection. When forest canopy is dry and
the surface appears smooth to the radar pulse, the radar energy may
be able to penetrate below the layer, giving rise to volume scattering.
Volume scattering may enhance or decrease image brightness,
depending on the amount of the returned energy. When radar look
direction is perpendicular to the orientation of linear features, such as
urban buildings, fences, mountain ranges, ocean waves, fault lines,
and row crops, a large portion of the radar energy will be reflected
back and the features will be bright.

Lidar technology provides detailed and accurate measurement of
surface elevation that only photogrammetry can emulate, but it does
not need to work with aerotriangulation. Most Lidar systems
record both the time of a return pulse and the intensity of the return.
Terrestrial topographic Lidar may be categorized into single-return,
multiple-return, and waveform Lidar. Single-return Lidar can be used
to derive a digital surface model. Multiple-return Lidar can be
employed for the derivation of 3D building and forest structure
information. The full waveform Lidar provides detailed information on vertical structure and accurate detection of the true ground

topography, and thus is very useful in forest analysis and habitat mapping. Two broad groups of Lidar applications are discussed in this chapter. Topographic mapping applications include digital surface model, digital elevation model, urban building and structures, transportation, utilities, and communications. In forestry studies, Lidar data have been used to measure forest parameters, such as stand density, tree height, crown width, and crown length, as well as to estimate forest volume, biomass, DBH and basal area. Lidar is a rapidly evolving technology, meaning that in the future we can find it in many more applications as this technology becomes more mature and data processing protocols become more defined.

Key Concepts and Terms

Active microwave systems   The remote sensing systems that provide their own source of microwave radiation to illuminate the target, such as radar.

Azimuth   Along-track dimension parallel to the flight direction in radar imaging.

Azimuth resolution   Radar image resolution in the azimuthal direction. It is determined by the angular width of the radar beam and slant range distance. The azimuth resolution becomes coarser outward from near to far range.

Corner reflection   High-rise buildings in urban areas may line up in such a way that the incoming radar pulses are able to bounce off the streets and then bounce again off the buildings, giving rise to a double reflection that yields strong return.

Cross polarization   The microwave energy that transmits to the target with horizontal orientation. The antenna receives vertically polarized backscattered energy, and vice versa.

Depression angle   Complementary to the look angle, it is defined as the angle between a horizontal plane and a given slant range direction. Depression angle becomes smaller when moving outward from near to far range.

Dielectric constant   A measure of both the conductivity and reflectivity of a material. It describes a material's capability to hold a charge and to polarize when subjected to an electric field.

Digital elevation model (DEM)   Also known as digital terrain model (DTM), it is a digital cartographic representations of the elevation of the Earth's surface at regularly spaced intervals in x and y directions, using z-values as a common vertical reference.

Digital surface model (DSM)   Similar to DEM or DTM, with the difference being that DSMs depict the elevation of the top surfaces of buildings, trees, and other features located above the bare Earth.

Discrete-return Lidar   A digitization of waveform Lidar. When the return waveform is above a certain threshold, the Lidar system records a discrete pulse—up to five discrete pulses per return. Lidar can be collected as a single return or multiple returns.

Far range   The portion of the swath farthest from the nadir.

Foreshortening   When the radar beam reaches the base of a tall feature tilted toward the radar (e.g., a mountain) before it reaches the top. The length of the slope will be reduced to an effective length of zero in slant range.

Ground range distance   The true horizontal distance along the ground corresponding to each point measured in slant range in radar imaging.

Ground range resolution   Radar image resolution in the range direction. The range resolution is dependent on pulse length and depression angle. For a fixed-slant range resolution, the ground range resolution varies inversely with the cosine of the depression angle, implying that ground range resolution becomes better outward from near to far range.

Horizontal polarization   The microwave energy that transmits to the target with horizontal orientation. The antenna receives horizontally polarized backscattered energy.

Incidence angle of radar   The angle between the radar beam and a line perpendicular to the surface. It increases as the radar beam moves across the swath from near to far range.

Layover   Occurs when the radar beam reaches the top of a tall feature before it reaches the base. The top of the feature is displaced toward the radar from its true position on the ground, and "lays over" the base of the feature.

Lidar   Acronym for light detection and ranging. Lidar is synonymous with airborne laser mapping and laser altimetry.

Look angle   The angle at which the radar "looks" at the surface.

Multiple-return Lidar   Current Lidar systems are capable of measuring multiple Lidar returns, up to five returns per pulse. A filter needs to be applied to detect peaks in the reflected waveform and to record the timing of those peaks as discrete "returns." These discrete Lidar returns can provide information about objects above the ground and the bare ground surface. Among all the returns, the first and last returns are the most important ones in topographic mapping. Multiple-return Lidar is required to assess 3D forest structure characteristics.

Near range   The portion of the image swath closest to the nadir.

Passive microwave systems   A passive microwave sensor detects the naturally emitted microwave energy within its field of view.

Radar   RAdio Detection And Ranging, an active remote sensing system that transmits a microwave signal toward the target, and detects the backscattered portion of the signal.

Radar image speckle   A grainy "salt and pepper" texture in radar images. Speckle is caused by random interference from the multiple scattering returns that will occur within each resolution cell.

Radar interference   Two radar waves with the same wavelength and frequency travel along in space but with slightly different starting points, which will result in phase differences.

Radar parallax   Apparent change in the position of an object due to an actual change in the point of view of observation. Parallax may be used to create stereo viewing of radar images.

Radar polarization   Refers to the orientation of the electric field. Radars are designed to transmit microwave radiation either horizontally polarized or vertically polarized.

Radar shadow   It occurs when the radar beam is not able to illuminate the ground surface. Both foreshortening and layover cause shadow.

Radargrammetry   The science of obtaining reliable measurements from stereo radar images. It is analogous to photogrammetry for stereo aerial photographs.

Range   The across-track dimension perpendicular to the flight direction in radar imaging.

Real aperture radar   A radar system in which the azimuth resolution is related to the length of the antenna and increasing beam width.

Single return Lidar   Early Lidar systems can record only one discrete return, either the first peak or the final peak of the reflected energy. Single return from the top of the vegetation canopy and buildings provides useful information. A digital surface model may be generated by interpolation of raw laser measurements.

Slant-range scale distortion   The distortion occurs because the radar is measuring the distance to features in slant-range rather than the true horizontal distance along the ground. This distortion results in a varying image scale, moving from near to far range, for example, causing targets in the near range to appear compressed relative to the far range.

Specular reflection   Flat surfaces, such as smooth water surfaces, airport runways, roads, and freeways, usually appear dark in radar images, reflecting little or no energy back to the radar.

Surface roughness   Subtle irregularities in the Earth's surface that are caused by textural features comparable in size to radar wavelength, such as leaves and twigs, closely spaced vegetation with various shapes, grass blades, pitted materials, granular soils, gravel, etc., and whose surfaces have dimensional variability on the order of millimeters to centimeters.

Synthetic aperture radar (SAR)   A synthetic aperture is constructed by moving a real aperture or antenna through a series of positions along the flight line. As the platform advances, SAR sends its pulses from different positions. At any instant in time, the transmitted beam propagates outward within a fan-shaped plane, perpendicular to the flight line, generating a "broad beam" through which targets remain to be illuminated for a longer period of time. All echoes from a target for each pulse are recorded during the entire time that the target is within the beam. By integrating the pulse echoes into a composite signal, the movement of the antenna helps to increase the effective length of the antenna, simulating a real aperture.

Vertical polarization   The microwave energy that transmits to the target with vertical orientation. The antenna receives vertically polarized back-scattered energy.

Volume scattering   Typically consists of multiple bounces and reflections from different components within the volume or medium. When the top layer of trees (forest canopy) is very dry and the surface appears smooth to the radar, the radar energy may be able to penetrate below the layer. In that case, radar backscatter may come from the canopy, the leaves and branches further below, and the tree trunks and soil at the ground level.

Waveform Lidar   The entire waveform from the laser pulse is recorded, as opposed to discrete returns. This technique captures more detailed information on vertical structure and can more accurately detect the true ground topography under dense vegetation canopy. In addition, it is not subject to signal processing errors. Currently, waveform Lidar has lower spatial resolution than discrete-return Lidar and is not widely available.

Review Questions

1.   What are some advantages and disadvantages for using radar images, compared to satellite images and aerial photographs?

2.   What are the major differences between real aperture radar and synthetic aperture radar?

3.   Justify the usefulness of radar imagery in military applications.

4.   As an active sensing technology, are there any localities where radar is more effective? Are there any places where radar is less effective?

5.   The geometry of radar imagery is very different from that of aerial photography and electro-optical sensing imagery. Why is knowledge of the geometry of radar data collection so important in understanding the spatial resolution of radar imagery and in interpreting the geometric characteristics of the imagery?

6. Radar images are particularly useful for studying topography (relief). Explain major strengths of radar imagery for such study, and discuss main problems (e.g., displacement errors) associated with the study of relief using radar images.

7. Explain how each of the transmission characteristics of radar signals (such as wavelength, polarization, and illumination angle) affect the collection of radar imagery.

8. Discuss the main factors that influence radar backscatter and thus image tone and texture. In interpreting radar imagery of a forest park, what are some of the unique surface characteristics that have to be considered?

9. What are major similarities and differences between Lidar and radar sensing?

10. Comparing with photogrammetry, what are some major advantages of Lidar technology? Are there any disadvantage?

11. Why is Lidar becoming a popular technique in urban planning and civic applications?

12. Forestry is a major application area of Lidar technology. Can you briefly explain the methods of this application?

13. From this chapter, you can see Lidar data are frequently used together with aerial photographs. Can you explain why this combined use is necessary? Can Lidar be used together with multispectral satellite imagery, such as Landsat TM data? Explain when such combination is necessary.

References

Arnold, R. H. 2004. *Interpretation of Airphotos and Remotely Sensed Imagery*. Long Grove, Ill.: Waveland Press.

Baltsavias, E. 1999. Airborne laser scanning: Basic relations and formulas. *ISPRS Journal of Photogrammetry and Remote Sensing*, 54:199–214.

Canada Centre for Remote Sensing. 2007. *Tutorial: Fundamentals of Remote Sensing*, http://www.ccrs.nrcan.gc.ca/resource/tutor/fundam/chapter2/08_e.php. Accessed on December 3, 2010.

Freeman, T. 1996. *What is Imaging Radar?* http://southport.jpl.nasa.gov/. Accessed on January 23, 2011.

Hecht, J. 1994. *Understanding Lasers: An Entry Level Guide*. New York: IEEE Press.

Hill, J. M., L. A. Graham, R. J. Henry, D. M. Cotter, A. Ding, and D. Young. 2000. Wide area topographic mapping and applications using airborne light detection and ranging (LIDAR) technology. *Photogrammetric Engineering & Remote Sensing*, 66(8):908–927.

Jensen, J. R. 2007. *Remote Sensing of the Environment: An Earth Resource Perspective*, 2nd ed. Upper Saddle River, N.J.: Prentice Hall.

Lefsky, M. A., W. B. Cohen, et al. 1999. Lidar remote sensing of the canopy structure and biophysical properties of Douglas-fir western hemlock forests. *Remote Sensing of Environment*, 70(3):339–361.

Lillesand, T. M., R. W. Kiefer, and J. W. Chipman. 2008. *Remote Sensing and Image Interpretation*, 6th ed. Hoboken, N.J.: John Wiley & Sons.

Miliaresis, G., and N. Kokkas. 2007. Segmentation and object-based classification for the extraction of the building class from LIDAR DEMs. *Computers & Geosciences*, 33(8):1076–1087.

Naesset, E. 2002. Predicting forest stand characteristics with airborne scanning laser using a practical two-stage procedure and field data. *Remote Sensing of Environment*, 80(1):88–99.

Popescu, S. C. 2011. Lidar remote sensing. In Weng, Q., Ed. *Advances in Environmental Remote Sensing: Sensors, Algorithms, and Applications*. Boca Raton, Fla.: CRC Press, pp. 57–84.

Popescu, S. C., R. H. Wynne and R. H. Nelson, 2003. Measuring individual tree crown diameter with LIDAR and assessing its influence on estimating forest volume and biomass. *Canadian Journal of Remote Sensing*, 29(5):564–577.

Popescu, S. C., and R. H. Wynne. 2004. Seeing the trees in the forest: Using Lidar and multispectral data fusion with local filtering and variable window size for estimating tree height. *Photogrammetric Engineering & Remote Sensing*, 70(5):589–604.

Sabins, F. F. 1997. *Remote Sensing: Principles and Interpretation*, 3rd ed. New York: W.H. Freeman and Company.

Short, N. M., Sr. 2009. *The Remote Sensing Tutorial*, http://rst.gsfc.nasa.gov/. Accessed January 25, 2011.

Webster, T. L., D. L. Forbes, S. Dickie, and R. Shreenan. 2004. Using topographic Lidar to map flood risk from storm-surge events for Charlottetown, Prince Edward Island, Canada. *Canadian Journal of Remote Sensing*, 30(1):64–76.

Index

Note: Page numbers followed by *f* denote figures; page numbers followed by *t* denote tables.